W9-AEV-278

Experimentation:

Prentice-Hall, Inc. *Englewood Cliffs, New Jersey*

An Introduction to Measurement Theory and Experiment Design

D. C. Baird *Associate Professor of Physics, Royal Military College of Canada*

EXPERIMENTATION: AN INTRODUCTION TO MEASUREMENT THEORY AND EXPERIMENT DESIGN. BY D. C. BAIRD.

First printing...........February, 1962
Second printing..........August, 1962
Third printing..........January, 1963
Fourth printing..........January, 1964
Fifth printing..............June, 1965

Library of Congress Catalog Card Number 62-11883

Printed in the United States of America 29534C

Preface

This text is intended for use in first and second year physics laboratory courses for scientists and engineers. The function of such courses varies widely, and this function is changing with time as new approaches to physics education are tried. However, regardless of the actual aim which the designer of a laboratory course has in mind, one point remains invariable. That is that experiments involve measurement. Unless the nature of measurement is clear to the experimenter, the benefit which is sought from carrying out the experiment cannot be fully realized. This text is written, therefore, on the premise that whatever the purpose being served by the laboratory course, consideration of measurement theory and elementary experiment design should not be entirely neglected. In any case, the emphasis on experiment design in engineering is growing to the point where an introduction at elementary levels is becoming essential, and this text is intended to contain the material necessary for this work.

For such a purpose, it is not enough merely to insist that physics laboratory students perform routine error calculations; the process as a whole has to make sense, and sufficient detail is included here to ensure that it does. The text is not intended to be a complete treatise on industrial statistics or on numerical methods in observation processing. It is merely an introduction, because the writer feels that it is necessary for the student to acquire, as a first step, a personal feeling for the nature of measurement and uncertainty, leaving mathematical sophistication to be acquired later. The amount of mathematical work has, therefore, been kept to the minimum which makes the treatment understandable. Examples have been added to some of the chapters because experience has shown that the average laboratory program does not provide the necessary amount or variety of exercise in the mechanics of observation processing.

The writer wishes to express his grateful thanks to all those who have helped in the assembling of this material. In particular he wishes to thank Dr. T. M. Brown and Mr. C. D. Pearse, the Royal Military College, who have provided invaluable assistance in discussing the principles on which it is based and Mr. A. J. Filmer, also of Royal Military College, who checked the answers to the problems.

D. C. B.

Contents

1 Introduction to Laboratory Work

It is the aim of this introduction to explain some of the purposes in physics laboratory work for the student of physics or engineering. So much time is spent either in the laboratory or on the work of the laboratory that value for the time spent will be gained only if the student understands clearly the purposes of the laboratory course and the means by which these purposes are being achieved. This understanding can best be found through a study of the nature of progress in science and technology, and the realization that one of the main steps in scientific or technological advancement is the experimental one. The nature of experiments will be discussed more completely in Chapter 3 and it will suffice for the moment merely to remind the reader of the vast significance of, and consequences from, experimental work. This gives observation a dominating role in almost every field of human activity. Consider, too, the enormous range of topic covered, from the well-defined, strictly controlled experiment of the physicist to the work of the biologist or physiologist, dominated to a large extent by statistical fluctuation, or the observations of the astronomer whose measurements may be precise but whose degree of control over his subject matter is somewhat limited. It is obvious, therefore, that it is useless to lay down a set of rules to say in general how experiments should be done: the range of activity is far too wide. However, there are habits of thought which will prove useful whether the observer is studying rabbits, stars, or tar in a road surface mix. It is the aim of this book to describe some of these principles and to show how the student may use introductory or secondary laboratory work to become practiced in such arts of experimental investigation as will be useful, no matter what line of work is ultimately to be pursued.

The important part of the approach is that the range of experimental technology is so vast that it is completely futile to attempt to prepare the student for every contingency. The chances are overwhelming that the graduate engineer or scientist will meet situations in his work which are new to him. Thus, one of the most important single factors in his education should be the development of the capacity to cope with new situations, and this book is a study in the methods whereby this can be accomplished.

Furthermore, professional research or technological development work is usually difficult through its very nature. This is natural, since the necessity for experimental work will always arise from the existence of a problem. Therefore, the future scientist or engineer must become accustomed to accepting practical work as a problem. He must gain practice in working at such problems until a way of solving them experimentally is found. He must, in addition, become capable of solving these problems on his own, since, unless he is capable of independent work, his worth to his employer is lessened. It must be clear that the necessary practice in solving problems experimentally is not gained by a laboratory system which provides the student with extensive and detailed instructions for the performance and calculation in the experiment. As will be seen later, the major part of experimental investigation is the preliminary analysis and the final evaluation of the experiment. If the experiment is properly planned, the actual performance can be reduced to a simple collecting of observations. Most of the thought goes before and afterwards. To permit the student to bypass these, the most important parts of the experiment, is to give him a totally wrong impression of the importance of the performance of the experiment and to deny him the opportunity of acquiring the real skill of experimenting. This capacity for independent thought is one of the most important single qualities of a research or development worker and he will acquire the necessary judgement only through the exercise of judgement.

This, then, suggests the attitude which the student should have towards his laboratory work. He should regard each experiment as a model of a problem which might be encountered in actual research or development. To be of any use at all the experiment must be a new situation and it must be a problem. Therefore, the student must expect to have to work at his laboratory problems, otherwise the time is wasted. He will be expected to make his own decisions about the method of making the measurements so as to achieve maximum efficiency of information collection. These decisions may often be wrong, but the student will learn more effectively in retrospect, and he will be strengthened by having made his own decisions. He will have to work within the framework of the apparatus provided, since he must learn that the skill in experimenting lies largely in achieving the maximum experimental yield with the resources available. Restrictions of time, too, merely simulate the conditions of most actual research or development. The experiments will never be ideal. However, this should not be regarded as a defect, but as a challenge. The real work of evaluating an experiment lies in sifting the information desired from the yield of the experiment, which is always clouded to a certain extent by uncertainty. The experimenter must learn to identify sources of uncertainty or error and, if possible, eliminate or allow for them. Whatever the degree of control he has over his experiment, however, he must evaluate the reliability of his result. This critical evaluation of the experiment is just as important as obtaining a numerical answer. The ability to cope with such conditions can be acquired only by experience and it is a common injustice to the student to leave him with the impression that the experiments are perfect.

It is necessary to keep an open mind towards experimenting and not allow an objectively analytical attitude to be hindered by a preconceived opinion of what "ought" to happen in the experiment. This open-minded spirit of investigation can be cramped by the setting down of a too-detailed "requirement" for the student to follow. The emphasis in the teaching laboratory should be on learning rather than doing.

Report writing should be approached in the same constructive spirit. The necessity for engineers and scientists to express themselves clearly and informatively through the written or spoken word has been frequently maintained, and is a matter of widespread concern. It is the writer's conviction that the acquiring of such fluency by scientists and engineers is as much the responsibility of the science departments as of the arts departments and the medium of such education is, to a very large extent, the laboratory report. Facility in clear and elegant exposition of scientific material can only be acquired through much practice, and this is the purpose in writing laboratory reports. This purpose will not be achieved unless the report writing is taken seriously as an opportunity to improve one's powers of exposition. A report which degenerates into a mere indication that the experiment has been performed is a waste of time to student and teacher alike. The student who takes pains with his reports deserves, and will profit from, careful, constructive discussion and criticism of his report; the writer regards such discussion of report and experiment as an essential feature of the system. The student will learn from his mistakes only after careful clarification.

Thus, in conclusion, we may say that the laboratory offers the student the opportunity to acquire many of the skills connected with the performance of his professional work. He will gain facility in the analysis of problems, the evaluation of the solution achieved, and the ability to describe his work clearly and informatively for the benefit of others. These qualities are among the most important tools of his trade and are well worth acquiring.

2 The Nature of Measurement

2.1 Measurement and Confidence

The observation of nature which constitutes an experiment will almost invariably take the form of a measurement. This is the case whether the experiment is of a precision type in which the answer is the magnitude of a quantity, or whether the measurements merely have the purpose of substantiating a qualitative conclusion. Because of this absolutely fundamental role of measurement it is necessary to consider in some detail what a measurement actually is. A true understanding of the nature of measurement would prevent many errors of interpretation which may be perpetrated at the conclusion level.

To answer the question—"What is a measurement?," it is easier to say what it is not. There are two classes of statement which can be made in human knowledge. The first is the kind exemplified by Euclid theorems which deal with humanly constructed definitions, and these are absolutely incontrovertible. If Euclid defines lines, planes, angles, etc., he can then say with absolute certainty that the sum of the angles in his triangle is 180°. This is exact truth because it is little more that a restatement of his own definitions. It is the kind of statement which a measurement is *not*. The other class of statement is concerned, not with precise, humanly constructed definitions, but with experience. In attempting to make a statement, not about our own ideas, but about the *external world*, we immediately lose the advantage of unequivocal exactness and lay ourselves open to all the frailties of human judgement. This, then, is the nature of a measurement; it is *a statement of the result of a human operation of observation*. It is not a matter of measuring the length of a box, getting a value of 6 in., and then saying "the length of this box *is* 6 in." This last statement can never be made (unless, of course, the Government introduces legislation to *define* the inch in terms of our box). In general the only possible statement that can be made is, "I have measured the length of this box and get a result of 6 in." This last statement illustrates the extent and limitations of statements made about the external world.

We have stressed the human and fallible nature of measurement and its consequently limited validity. This leads to a second concept. The days are past when people made measurements for their own satisfaction. Today we are almost invariably under the necessity of conveying our experimental results to someone else, either the man in charge of our work, or else other workers in the field. This necessity immediately raises the question of confidence. Quite plainly—are people going to believe your measurement or not? It is clear that only your professional reputation can instill confidence in other people but some measure of the reliability of your measurement is obviously called for. Convinced of the subjective nature of measurement, we can ask, on being told that an experimenter has measured the box and produced an answer of 6 in.: Is the observer a metrologist who means that he has an answer of 6.000000 in., or was it measured by the local drunk who had lost his glasses and really meant that he thought it was somewhere between 5 and 7 in.? It should be noted here that such concern tends to arise only when the measurement reaches beyond the bounds of the immediately familiar. If a painter whom I shall employ to paint my house asks how long it is, I say "30 feet" and he is happy, for he knows in general terms how much paint to order. But if someone at a nuclear reactor research establishment has measured the thermal expansion coefficient of a new alloy for cladding reactor fuel elements, those who are going to use his measurements in new reactor design will inquire very carefully into the reliability of his stated result.

If, therefore, we wish to convince other people of the usefulness of our experimental result, the statement of the result must be amplified by the quotation of some range of confidence. The intelligent way to quote the answer would be, "I find the length of the box to be 5.95 in. with 95 per cent confidence that it lies between 5.90 and 6.00 in." Note that this still is merely an expression of opinion by the observer and if real confidence is to be justified, sufficient description of the mode of measurement must be given to allow the reader to form his own judgement of the value of the measurement. Our inebriated friend without his glasses might use his imagination and claim a measurement of 5.8279436 in., but no one would believe him. Moreover, there have, in fact, been many instances in physics where work, even by very distinguished physicists, has been shown later to contain errors much larger than the limits of uncertainty quoted by the original authors. The only thing one can do is to give as complete a description of the experiment as is feasible and, relying on your own professional reputation, give your suggestion as carefully as possible for the limits of confidence. The remainder of this work will be largely concerned with the methods available to do this.

It can be said that the confidence one has in a measurement and the closeness of the limits set for the uncertainty are inversely related. For example, I have a right to be as certain as any human being can be that the length of this room is between 1 m and 10 m but my faith in the statement that it lies between 5.478339 m and 5.478340 m is limited. In connection with this topic it is of interest at this point to note another aspect of the statement that confidence and precision are inversely related. It is commonly said that there is a distinction between observations permitting a unique, exact answer (normally of a counting nature), such as the number of people in this room, and those which are subject to uncertainty, such as the length of the room. However, the distinction is one of degree only, because most physical measurements have initially the character of a counting process (I can lay down my meter stick 5 complete times in succession along the wall of the room) and the uncertainties (and declining confidence) appear only when one tries to specify the limits more closely (is the length between 5.4 and 5.5 m? Or, is it, even more unlikely, between 5.4783 m and 5.4784 m?). This problem does not arise in many counting type experiments such as that mentioned above because they deal essentially with units only: one does not subdivide people.

2.2 Types of Uncertainty

The methods by which the uncertainty can be estimated do not depend on particular sources of discrepancy but it will be instructive before proceeding to clarify our ideas by reference to a few common situations. The sources of perturbation of measurements cover such a wide range that it is possible to do no more than list a few typical headings. It is common to divide such perturbations into categories such as random, systematic, personal, etc. However, these terms are frequently too precise for the average experimental situation and their usefulness is consequently limited. They are commonly defined as follows:

Random error is said to be shown when repeated measurements of the same quantity give rise to differing values.

Systematic error refers to a perturbation which influences all measurements of a particular quantity equally.

However, these terms must be used with caution since a set of readings will show truly random error only if there are a large number of small perturbing influences. If the discrepancies arise from only a few types of experimental defect, an analysis into a few competing systematic errors may be possible and, even more important, the statistical theory to be described later will not be applicable. On the other hand, an error which is systematic under one system of measurement (e.g., a set of ammeter readings all taken going the same way when the meter bearings are sticky) may become apparently random if the mode of measurement is changed (e.g., if the meter readings are taken with arbitrary current changes). We shall use the terms systematic and random to indicate only clear-cut cases.

As has been stated above, any measurement is subject to an enormous range of sources of perturbation. It is impossible to consider all of these as systematic errors and even the skilled experimenter will probably give up and leave a residuum of uncertainty to be treated under the heading of "random" error.

A few factors of common occurrence which limit the precision of a measurement are the following:

(a) *Instrument Calibration*

Clearly this is of paramount importance, for the observer is dependent on his scale for the accuracy of his measurement. His meter stick might have warped, his stop watch changed its rate, his ammeter be sensitive to position. This type of error is usually so regular as to justify the term "systematic." Only comparison with a standard instrument will enable this source of uncertainty to be removed or allowed for, and such comparison should be contemplated wherever a measurement plays a dominating role in an experiment. Those instruments which have built-in standards for comparison purposes (e.g., most direct reading potentiometers) are a great convenience and give corresponding confidence in the measurement.

(b) *Instrument Reproducibility*

Even when the over-all calibration of an instrument has been checked under a certain set of circumstances, mechanical defects can still influence the readings and can remove the value of the calibration unless the subsequent readings are taken in exactly the same fashion as the calibration values. Such defects are slackness and friction in meter bearings, backlash in micrometer measurements, and many other cases where the final meter indication is dependent on the path by which it is reached. Depending on the measurement schedule, this error may be systematic under very regular circumstances (even zero under some circumstances, such as a travelling microscope with all the readings taken going one way) or otherwise the deviations may be scattered sufficiently to justify the term "random." Many instruments such as precision ammeters will have a value marked on them for the magnitude of this uncertainty.

(c) *Observer Skill*

Even in this day of automation in large-scale research and development, the operation of scientific measurement is, to a very large extent, based on personal manipulation of the apparatus and visual reading of a resultant indication. Judgement will, therefore, enter at a number of points depending on the number of variables on which the result is dependent. This, consequently, magnifies the influence of uncertainties in such settings and has the result that the precision of an observer's measurement may be much less than that suggested by his ability to read the final answer on the scale. As an elementary example of this, consider the observation of the position of a spectrum line on a spectrometer. Here the preliminary manipulation consists of setting the spectrometer telescope cross wires on the spectrum line, and the measurement is a simple reading on a vernier scale. If the visual conditions are poor, the difficulty in the preliminary setting may be so great as to exercise a dominant influence over the over-all precision of the measurement. It must be stressed that the influence of the observer's judgement extends very far beyond the mere scale reading, and it is this integrated uncertainty which must be considered. Uncertainties of personal origin may be systematic, as in the case of the astronomer watching a star in a transit circle and consistently pushing the chronograph button late, or may be so varied in composition as to justify the term "random."

(d) *Miscellaneous Errors*

In an experiment involving more than one or two variables or factors which influence the final measurement, there are bound to be perturbations which influence the final reading. These perturbations are usually random in nature, such as line voltage fluctuations, vibration of instrument supports, variable cosmic ray background, etc. It is part of the task of the experimenter to reduce the influence of these perturbations to the minimum, but there will usually be a residual contribution.

(e) *Fineness of Scale Division*

Even with an ideally calibrated instrument under ideal conditions a fundamental limit is set to the precision of the measurement by the instrument scale, which is necessarily subdivided at finite intervals. Note that this is similar to the circumstances in which a numerical value is rounded off to some particular number of significant figures. This rounding off is equivalent to a statement that, whatever the actual value of the uncertainty, it cannot be less than 5 in the next decimal place. Thus, if we quote π as 3.1416 we are conveying no more information than that π lies between 3.14155 and 3.14165 and so the range of uncertainty of the statement is $\pm .00005$ about 3.1416. This is not a real statistical uncertainty, but just another way of saying, "the value of π is stated to 4 places of decimals." Reading an instrument like a meter stick to millimeters is just a form of rounding off and consequently carries an immediate implication of a minimum uncertainty of one-half the finest scale division. The possibility exists, of course, of visual interpolation between the finest scale divisions, but the validity of such a procedure depends largely on the size of the undivided interval. A claim to be able to interpolate within millimeter divisions would be difficult to justify, but if the scale were divided in centimeters a practised observer could probably judge millimeters quite reliably. In all cases of visual interpolation, however, the estimation of the actual reliability is very difficult.

The foregoing are merely a few of the headings under which experimental uncertainties can be grouped. Sometimes it is possible to assign a cause to a particular observed uncertainty, and sometimes the cause can be eliminated. Certainly all justifiable effort should be expended on such diagnosis and cure. However, there is always a limit to the resources of time or equipment available for such attempts, and there comes a stage at which the observer wishes to state the residual uncertainty in his reading. It is with this residual uncertainty that the following treatment is concerned. If the observer's measurement is to be regarded as significant, he must be realistic in his assessment of the uncertainty. If he claims too high a precision, he may be guilty of dishonesty and he will certainly cause grief to those who use his results for continued work. If he claims a pessimistically low precision, there may not be the serious consequences of the other extreme, but he is certainly doing himself an injustice. He is also degrading the worth of his work since it will not be accorded the significance which may be its due. This is unfortunate, not only for the experimenter himself, but also for the workers in the field who are denied confidence in results which could be of value. The point is, therefore, that pains should be taken to assess the uncertainty of the measurement, as carefully as possible, striving to be realistic without undue optimism or pessimism.

It is now necessary to face the problem of how such an estimate of precision can actually be made. We are, for the moment, considering only the uncertainty of a single measured value, and the over-all precision of an experiment will be considered later. If an observer makes one, and only one, measurement of a quantity, the problem of estimating its precision is a difficult one. It is not an uncommon suggestion to use the finest scale division as a measure of the "maximum" range of uncertainty. However, this practice does not take into account the possibility of fluctuation arising from other sources. It is foolish to claim limits of uncertainty of $\pm\frac{1}{2}$ mm on a meter stick if the influence of parallax could give a range of 2 or 3 mm, or to claim that voltages read on a meter divided in $\frac{1}{10}$ v are significant to $\pm.05$ v if random perturbations are causing the needle to fluctuate over 2 or 3 tenths of a volt. On the other hand, the manipulation of the apparatus could be sufficiently precise that settings can be made well within the limits of the scale subdivision. In this case interpolation between the finest scale divisions may become significant and the use of the scale division as a measure of the precision would be unnecessarily pessimistic. It is therefore of rather limited value merely to quote the scale division as the limit of uncertainty. If such a criterion is used, the nature of the uncertainty quoted should be clearly stated so that the significance of the estimate can be evaluated. This method of quoting uncertainty may possibly be justified by convenience, but only in the case of the most simple, direct measurements. In the case of most experimental observations such an estimate of precision is hardly of any more value than no estimate at all, for it can never be anything more than an estimate and we shall always require an actual measurement of the uncertainty. Thus, for an experiment of any significance, a single measurement of any quantity is almost useless. The only way of meeting this difficulty is to repeat the measurement, not so much with a mystical faith that the averaging results in a "better" value, as to see how closely reproducible the measurement is.

2.3 Distribution Curves

The process of duplicating readings is at the heart of any experimental process and much of the treatment to be given later is concerned with the ways in which this duplication can be carried out so that the yield of the experiment can be maximized.

In the meantime we shall concern ourselves solely with the consequences of straightforward duplication of the same measurement. Since we are dealing with a system which is, by hypothesis, subject to perturbation, the resulting readings will not be the same. Depending on the degree of perturbation or of judgement required, the spread may be a large or small fraction of the magnitude of the measurement. If the spread is large compared with the absolute value of the measured quantity, we can say, in general terms, that the reliability or precision of the measurements is poor, and vice versa. However, the problem is not restricted merely to making the observations. If the measurement is to be of any use in further work, or to other people, it must be capable of being described in simple terms. The simplest thing is merely to reproduce your table of results saying, "I have measured this quantity n times and here are the n answers." This places the burden of evaluating the significance of the results on the reader. However, such a practice, although convenient to the experimenter, is clumsy and obscure, since few people can see clearly the essential characteristics of a set of numbers just by looking at them. Thus, it is an enormous convenience and source of clarification if the way in which the readings are distributed along the range of possible values is apparent, and a visual representation is much the best way. Methods of picturing ranges of numerical values can take many forms, but the most profitable device is the "histogram."

This diagram is drawn by dividing the original set of observations into intervals of predetermined magnitude and counting the numbers of observations found within each interval. If one plots, on a suitable scale, this frequency versus the readings themselves, a block diagram is obtained which is the required picture of the distribution of the readings along the scale of values. It actually contains no more information than the original set of readings but has, as required, the enormous advantage of visual presentation of the nature of the distribution of the readings.

Such a histogram and the set of readings from which it is derived are shown in Fig. 2.1. If the number of readings is very high, so that a fine subdivision of the scale of values can be made, the histogram approaches a continuous curve, and this is called a *distribution curve*.

Clearly a histogram or a distribution curve contains the information about the spread of the experimenter's read-

Table 1

85	109	114	121	127	131
92	109	114	121	127	132
96	110	114	122	127	133
97	110	115	122	127	134
97	111	116	122	128	134
97	111	116	122	128	134
100	111	116	122	128	134
101	111	117	123	128	135
101	111	117	123	128	136
102	112	118	123	128	137
102	112	118	123	130	137
103	112	119	123	130	137
103	113	119	124	130	144
105	113	120	124	130	148
106	113	120	124	130	149
106	113	120	125	130	
107	113	120	125	131	
108	113	121	125	131	
108	114	121	126	131	

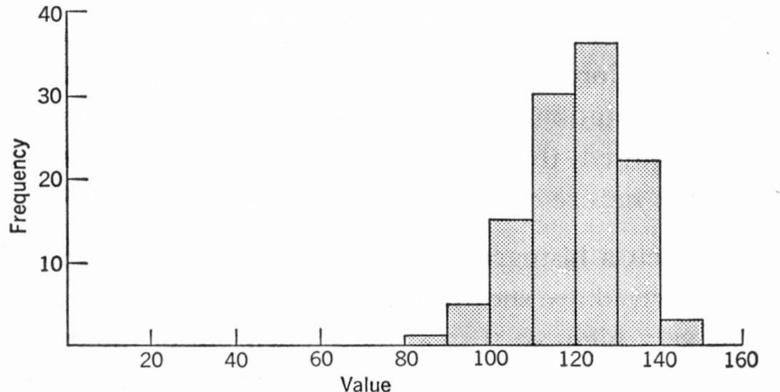

Fig. 2.1 A set of observations and its histogram.

ings which he wishes to communicate to other people, and does so in a clearer and more informative manner than does the original table of values. The histogram is an excellent way of presenting experimental results and is commonly used in reports on work which shows a high level of statistical fluctuation. It has the further advantage of presenting the observations themselves, free from manipulation on the part of the experimenter, and thus permits the reader to form his own judgement regarding the value of the work.

However, such a full-scale representation of the results may be undesirable, perhaps on grounds of difficulty of presentation, or more significantly, the results may be required for further work. In either case, one wants to find numbers which will represent the distribution of the values so as to answer the questions, "Which single number shall I take as the answer, and how reliable is it?" It is necessary, therefore, to define quantities which will serve as a "best" value and an "uncertainty."

It is difficult to say at this stage exactly what we mean by "best" and "uncertainty" because we are, for the moment, using a frankly intuitive approach. The uncertainty is obviously associated in some way with the spread of the results, and the best value with any tendency of the results to cluster in the middle of the distribution. As the development of the uncertainty theory progresses our concept of the terms above will gradually become clearer and clearer. If these numbers are to have general significance, they must be defined in some standard way so that their significance will be widely understood and accepted. Unfortunately, the significance of most numbers one can suggest for the purpose above depends on the actual shape of the distribution curve. We have not yet said anything about the shape of the curve itself and, clearly, this depends on the nature of the measurement. Consider an experiment which consists of making a measurement on a scale marked in inches and attempting to estimate tenths. If the most commonly found value were roughly halfway between divisions, it might be found that the probability of getting a reading declined to zero at the two boundaries and was approximately constant in the center third of the range. The distribution curve for such a set of readings might appear as in Fig. 2.2. On the other hand, the experiment might consist of measuring the tensile strength of a set of supposedly identical wire samples. Here repetition might show that, although there was a sharp upper limit to the tensile strength obtained, many specimens showed lower strength because of imperfections. This distribution curve might appear as in Fig. 2.3. The characteristics of these two curves are completely different, and it is obviously impossible to

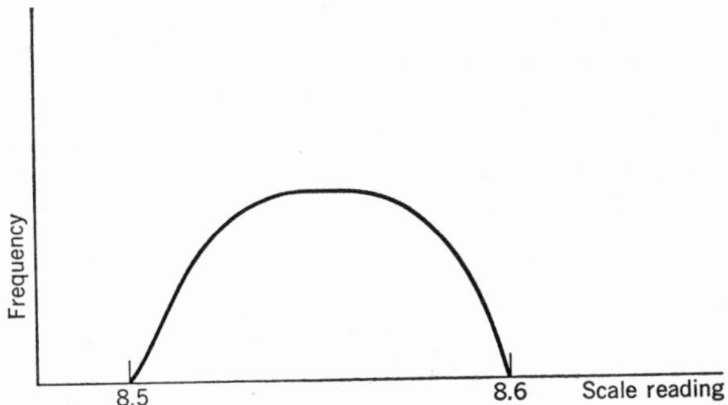

Fig 2.2 One type of distribution curve.

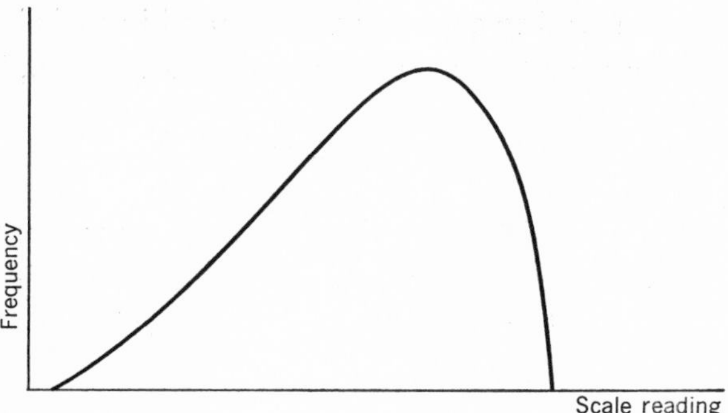

Fig. 2.3 Another type of distribution curve.

quote numbers which will represent them with any reliability.

The circumventing of this problem takes us into the domain of mathematical statistical theory. The principle is that, if any arbitrarily found distribution curve is too difficult to work with, we must define one which will prove tractable and, we hope, not too unrealistic. It must be stressed that most of the material in the rest of this chapter refers to this defined distribution curve only. If the statistical theory is to be of any significance in a set of measurements, the correspondence between the theoretical form of the curve and the distribution curve of the actual observations must be established, or else only limited validity of the statistical theory must be accepted and admitted.

Before working with the formal curve, however, there are some definitions relating to the distribution curves which are made independently of the shape of the curve. It must be remembered, however, that the significance of the quantities so defined does depend on the particular distri-

The square root of the variance is called the standard deviation, denoted by s, so that*

$$s = \sqrt{\Sigma\,[(x_i - \bar{x})^2]/n} \qquad (2.3)$$

Thus it may suffice, when stating the result of a repeated observation, to quote the mean value as the best value, and this standard deviation for the set of readings as a measure of the uncertainty. (For a slight correction to this statement see page 35.) It must be stressed again, however, that the significance of these quantities depends on the actual shape of the distribution curve. If for any reason the curve of the actual observations is markedly asymmetric, considerable care must be exercised in the interpretation of these calculated quantities.

It has been common in the past to define a quantity known as the *probable error*. This is a value which divides the area under the distribution curve into two equal parts denoted I and II in Fig. 2.5. It has the significance that any reading of the set has an equal probability of being inside and outside the limits set by the probable error. This is a useful and reasonable definition, but the size of the probable error, and its relation to the standard deviation depend on the particular shape of the distribution curve. This limits the usefulness of the quantity and it has become more common to disregard it in favor of the standard deviation.

The problem is thus a matter of reducing a set of observations to such a condensed form as will permit further work or calculation. The nature of this condensed form depends on the nature of the results. If the distribution

* Note that this formula involves the sum of the squares of the deviations, *not* the square of the sum of the deviations.

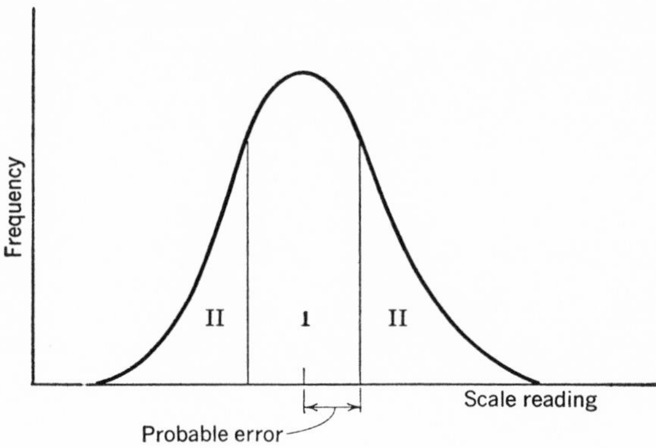

2.5 The probable error in a frequency distribution.

curve is symmetric, the obvious quantities to quote are the central value (i.e., the mean) and some measure of the width of the curve, such as the standard deviation of the set. If the distribution curve is markedly asymmetric, it may be necessary, in order to provide an accurate impression of the results, to quote the mode and/or median, in addition to the mean, and, in extreme cases, there is no alternative to quoting the whole distribution curve.

Two difficulties remain for the physicist. First, the distribution curve for a particular measurement is not always available. If one is measuring a wire diameter with a micrometer, one rarely takes the hundreds of measurements required to provide a significant distribution curve. Second, it would be useful if the standard deviation could have a definite numerical significance as a measure of the uncertainty for, as we have stated above, the relationship of a quantity such as the standard deviation to the distribution curve as a whole depends on the shape of the curve. For this purpose it is common in physical measurement not to consider the actual distribution curve relating to a particular measurement, but to discuss the situation in relation to a defined curve. We do not pretend that all physical observations actually follow this curve but many carefully made observations may be adequately close. Furthermore the procedure will assign a definite numerical significance to quantities such as the standard deviation and also permit deductive work like the theory of sampling to have a definite numerical significance.

2.4 The Gaussian or Normal Distribution

Many theoretical distribution curves have been defined and their properties evaluated, but the one which has most significance in the theory of measurement is the Gaussian distribution.

The Gaussian error curve can be defined on the assumption that the total deviation of a measured quantity, x, from a central value, X, is the consequence of a large number of small fluctuations. If there are m such contributions to the total deviation, each of equal magnitude, a, and either positive or negative, the total set of observations may range from $X + ma$, if all fluctuations happen to be positive simultaneously, to $X - ma$ if the same happens in the negative direction. It can be shown, in such a random summation of positive and negative quantities (as in the "random walk"), that the most probable sum is zero, meaning that the most common values of x are in the vicinity of X. The distribution curve is thus peaked in the middle, is symmetric, and declines smoothly to zero at $x = X + ma$ and $x = X - ma$. If this concept is taken to the limiting case of an infinite number of infinitesimal contributions to the total deviation, the curve has the form shown in Fig. 2.6. Treating the curve solely from the mathematical point of view for the moment, its equation can be written

$$y = C\, e^{-h^2(x-X)^2} \tag{2.4}$$

Here the constant C is a measure of the height scale, since $y = C$ for $x = X$ at the center of the distribution. The curve is symmetric about $x = X$ and approaches zero asymptotically. The quantity h obviously governs the

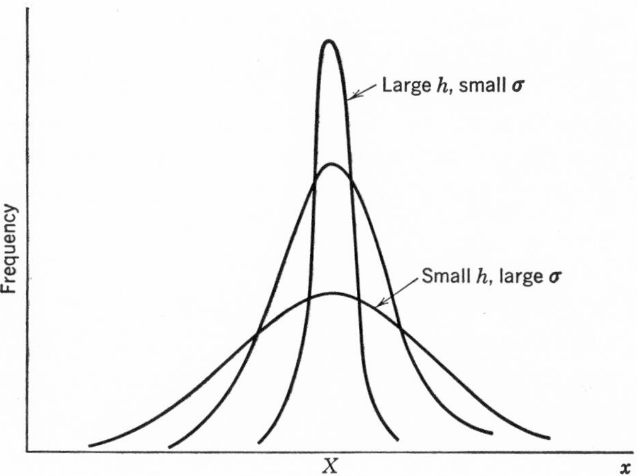

Fig. 2.6 The Gaussian distribution curve.

width of the curve, since it is only a multiplier on the x scale. If h is large, the curve is narrow and high; if small, the curve is low and broad. The quantity h must clearly be connected with the standard deviation, σ, of the distribution and it can be shown that the relationship is

$$\sigma = \frac{1}{\sqrt{2}\, h} \tag{2.5}$$

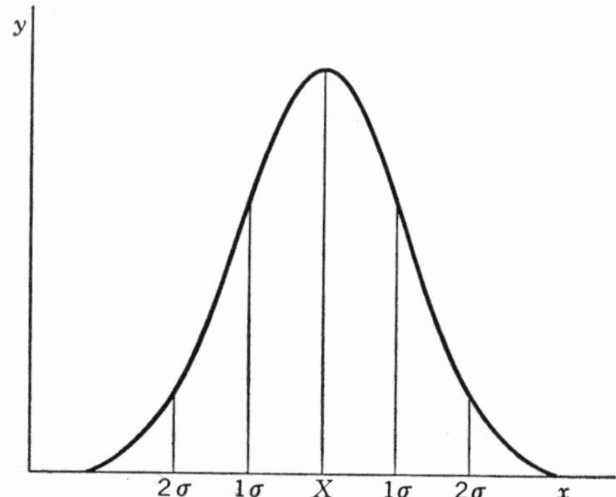

Fig. 2.7 The relationship of 1σ and 2σ limits to the Gaussian distribution.

(We shall use latin letters, e.g., s for standard deviation, for quantities related to finite sets of actual observations, and greek letters, e.g., σ, when referring to the normal distribution itself, or a "universe" of readings as described on page 30). The relationship of the standard deviation to the scale of the curve is indicated in Fig. 2.7 by the lines drawn at intervals of 1σ and 2σ from the central value. The probable error can also be evaluated and it is given by

$$p = \frac{0.48}{h} \approx 0.67\sigma \tag{2.6}$$

For a more complete account of the mathematical properties of the Gaussian error curve see Appendix 1.

2.5 Correspondence between the Normal Distribution and Actual Observations

The normal distribution is a smooth and continuous curve. It can, therefore, correspond only with a limiting case as the number of observations tends to infinity, and this rather stringent requirement is the first point of difference between the Gaussian curve and actual sets of observations. Even when a numerous set of observations is made, however, there is no guarantee that the distribution will be Gaussian. One of the assumptions in the definition of the Gaussian distribution is that, in each reading, there is an infinite number of infinitesimal perturbations, each of which is equally likely to be positive or negative. This is an idealized condition, and the coarser perturbing influences in actual observation make it rather difficult to establish correspondence between actual observations and the ideal form of the curve. For example, one result of this is that the Gaussian curve approaches the axis only asymptotically, thus permitting a finite possibility of extremely far-out readings. This probability is certainly small but practical considerations make such possibilities completely unacceptable.

Some forms of observation may not even accord in principle with the normal distribution. One common example is particle counting in nuclear physics. Here the repetition of counts, over equal time intervals, of randomly occurring events follows a skew distribution known as the *Poisson distribution.* If the number of counts in each interval is small the distribution is markedly skew, but the curve approximates the normal distribution more closely as the number of counts increases. The statistics of this distribution play a very important role in nuclear physics but it is not discussed here further because its application is limited to counting experiments. For a description of the distribution and the ways of using it the reader is referred to Reference 18 in the Bibliography.

2.6 Significance of the Standard Deviation in Actual Measurements

From Fig. 2.7 it can be seen that the values 1σ, 2σ, etc. divide the area under the curve into various regions. Since areas under a frequency distribution curve represent numbers of readings, areas, expressed as a fraction of the total area, are a measure of the probability of obtaining readings within the range defining the area. It is now possible to see the significance of the curve and the standard deviation to the experimenter.

Imagine that the night before an experimenter is going to make a measurement of, say, the period of a pendulum, the ghost of Galileo comes to the laboratory and, with the same apparatus and under the same conditions to be used in the experiment, makes a very large number of observations (say a million) of the pendulum's period. He can then draw the distribution curve of these and this distribution curve can, for the present purpose, be assumed to be very closely Gaussian. It will be centered on a mean value X which is very close indeed to the value obtained if the number of observations actually went to infinity. This last limit would define (ignoring systematic error) what we could term the "true" value. There will also be a certain value for the standard deviation which can be calculated from the readings using Equation (2.3). This picture is merely a way of visualizing what is called the "universe" or "population" of a particular reading. This term refers to the infinite set of readings which could be made with the apparatus, and thus provides a link between actual observations and the statistical theory. This population and its distribution curve exist for any measurement we care to consider and it is the constants of this curve, particularly the mean value X, which is the object of the experimenter's work. The only trouble is its obvious inaccessibility.

Next morning our experimenter arrives and makes his single measurement. It will fall somewhere along the range of the previously obtained million readings. Its exact position is a matter of chance, and all that can be said is that the probability of its being within a range of $\pm 1\sigma$ from the value of X is the fraction that the area enclosed by the 1σ limits is of the total area under the curve. This can be shown (see Appendix 1) to be about 68 per cent for a Gaussian distribution. The probability that it falls within 2σ is, correspondingly, about 95 per cent. Thus, the only thing that can be stated about our experimenter's measurement is that he stands a 68 per cent chance of being within $\pm 1\sigma$ of Galileo's almost "true" value of X and a 95 per cent chance of being within $\pm 2\sigma$. The only trouble is that, in the absence of psychic contact with Galileo, he does not know what values of X and σ are appropriate to the experiment. All he can say is that his single reading stands a 68 per cent chance of being within something of something—a somewhat restricted statement. This brings us back to the earlier point of the futility of isolated readings. The obvious solution, which has already been propounded, is the duplication of readings in an attempt to increase the amount of information available, and this requires an extension of the statistical theory.

2.7 Sampling

The two unknowns which constitute our experimenter's problem are X, effectively the "true" value, and σ. Obviously, if the experimenter could make another million readings of his own, he would be able to get sufficiently close to X from the peak of his distribution curve or the mean of his observations. In this case he need not necessarily be interested in the standard deviation since he would have obtained his answer sufficiently precisely. The trouble is that obvious practical considerations preclude the taking of another million readings. He will attempt to duplicate his readings, of course, but the best he will achieve will be a *sample* of the next million. Say he takes 10 readings. He hopes that this will do two things. First, it will give him some measure of σ, and second, he trusts that he will be further rewarded for his pains by improved reliability of his mean value, although he cannot hope that the mean of his sample should actually coincide with Galileo's mean of a million readings. The improvement that does result is given by the statistical theory of sampling.

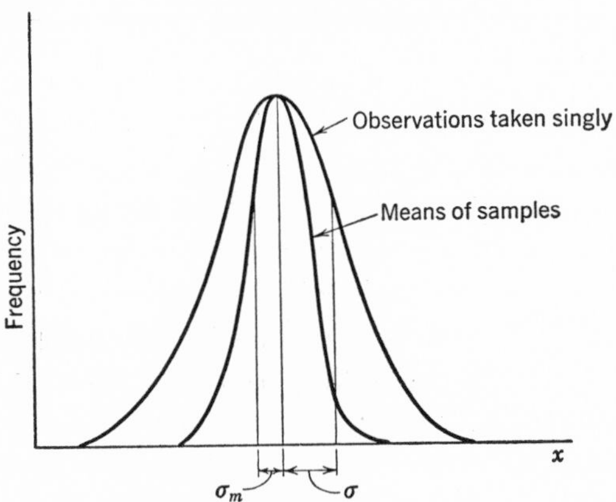

Fig. 2.8 Distribution curve of single observations and sample means. (Note that the vertical scale for the two curves is not the same. They have been plotted with a common peak value solely for purposes of illustration.)

Suppose that Galileo returns again the next night and takes another million readings. Let us suppose that he then takes these million readings and divides them into groups of ten. He will have a hundred thousand of these. Suppose that he works out the mean of each group and plots the distribution curve of these means. A derivation based on the statistics of sampling shows that this new distribution curve of means will also be Gaussian (even if, in fact, the distribution of single observations was not Gaussian) and will be centered on the same value X as was the first curve. However, its most striking feature is that, as illustrated in Fig. 2.8, it is narrower than the distribution curve of the readings taken singly and it can be shown that the standard deviation of this set of means, which we shall call σ_m, is given by

$$\sigma_m = \frac{\sigma}{\sqrt{n}} \tag{2.7}$$

where n is the number of readings in each group of the complete set. (For an indication of how this can be proved see page 64.) In the case of our illustration n equals 10, and so Galileo will find the distribution curve of his collection of means to be about one-third as wide as that of the readings taken singly.

Hence, the improvement which the experimenter hoped to gain from duplicating his readings comes from this smaller standard deviation of the means of the groups. The point is that his action of taking a set of 10 readings is just that of selecting, at random, one of Galileo's groups. If, therefore, he evaluates the mean of his sample, the likely position of this quantity along the scale of values is governed by the

distribution curve of means, and not that of the values taken singly. Thus, if he takes n readings and evaluates their mean, he stands a 68 per cent chance that this mean will come within, not $\pm\sigma$, but $\pm\sigma/\sqrt{n}$ of Galileo's central value X and, correspondingly a 95 per cent chance of coming within $\pm2\sigma/\sqrt{n}$.

This does justify his faith that duplicating his readings will yield him a better answer since, with 10 readings he is likely to be 3 times better off than before. However, even with this improvement, notice that, in the absence of the value of σ, he is still in the position of being able to say nothing more than that he stands a 68 per cent chance of being within something of something. He can improve his situation only by making an estimate of σ. Towards this, all he can do is work out the s for his sample. To see what luck he will have, let us suppose that Galileo goes back to his hundred thousand groups of readings and, for each group, works out its s value. They will, of course, not all be the same, and, again, an extension of the mathematical treatment of the original Gaussian equation shows that this set of s values will form a distribution curve which is, as before, bell-shaped and which is centered on σ as shown in Fig. 2.9. The breadth of this curve is related to the standard deviation of the set of s values. We shall call this σ_s and it can be shown that it is given by

$$\sigma_s = \frac{\sigma}{\sqrt{2(n-1)}} \qquad (2.8)$$

where n is the number of readings in the sample. Thus, if our observer with his sample of 10 observations evaluates the s value of his sample, he will get a value which will lie somewhere along the scale of possible s values, and this is

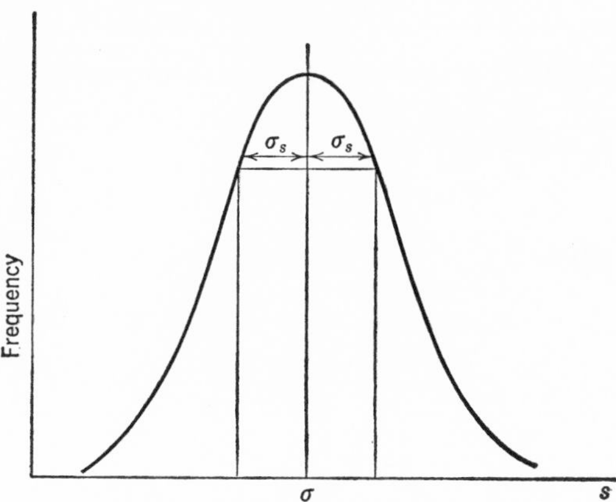

Fig. 2.9 The distribution of sample standard deviations.

as far as will be able to go towards a value of σ. It can be shown, actually, that the best estimate of σ that can be made from a sample is not exactly the sample standard deviation itself. Instead, the best estimate of σ is given by

$$\text{best estimate of } \sigma = \sqrt{\Sigma[(x_i - \bar{x})^2]/(n-1)} \qquad (2.9)$$

with $n - 1$ in the denominator instead of n. We shall accept this value and when we quote the standard deviation for a set of readings in the future we shall do so, not as a measure of the variance of the set itself, but rather as the best estimate of the universe standard deviation σ. The difference is, in any case, small if we have enough readings to justify a calculation of s.

This estimate of σ from the sample is the best that the experimenter can do but he can now state limits to his ignorance, for Equation (2.8) shows that, with 10 readings, he stands a 68 per cent chance that his s value will lie within a range of $\pm\sigma/\sqrt{18}$ or, approximately $\pm\sigma/4$, about the unattainable and elusive σ. He has correspondingly a 95 per cent chance of coming within $\pm\sigma/2$ approximately.

Let us list some typical values of $\sqrt{2(n-1)}$:

Table 2

Confidence 68%		Confidence 95%	
n	$\sqrt{2(n-1)}$	n	$\frac{1}{2}\sqrt{2(n-1)}$
2	1.4	2	0.7
3	2.0	3	1.0
4	2.4	4	1.2
5	2.8	5	1.4
6	3.1	6	1.6
7	3.4	7	1.7
8	3.7	8	1.8
9	4.0	9	2.0
10	4.2	10	2.1
15	5.2	15	2.6
20	6.1	20	3.2
50	9.8	50	4.9
100	14.1	100	7.0

These values are illustrated in Fig. 2.10 for $n = 10$ and $n = 3$, giving an impression of the range of possible s values. On these curves are marked the $1\sigma_s$ limits, showing the range within which the observer stands a 68 per cent chance of having his sample standard deviation fall. The situation looks correspondingly worse for the $2\sigma_s$ limits and it can be seen that for $n = 10$, 95 per cent of the possible s values lie within limits of $\pm\sigma/2$, and for $n = 3$, 95 per cent lie within a range which is large enough to cover 0 up to 2σ. In actual practice the value of σ is, of course, un-

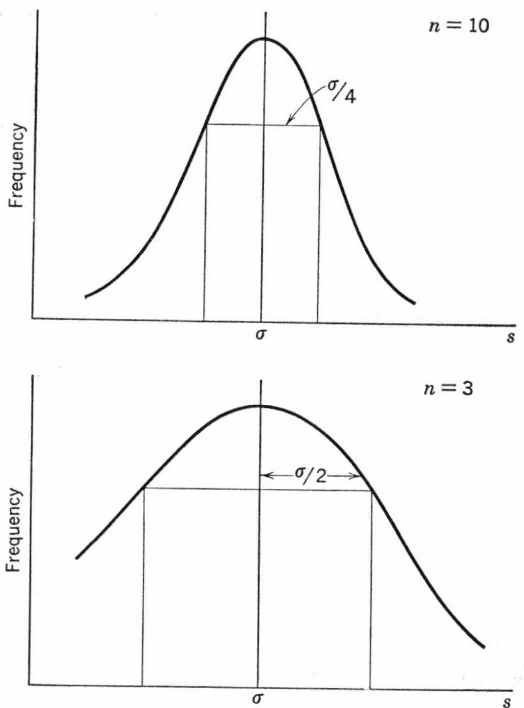

Fig. 2.10 Sample standard deviation distributions for small samples.

known. One would therefore use the value for the best estimate of σ in Equation (2.8) to give the actual numerical value for σ_s. This provides the actual range within which we have 68 per cent confidence that the standard deviation estimate lies.

The important point about this result is that the observer may be guilty of misrepresentation in quoting his results if he does not state his number of readings. For, if he quotes his mean value and the standard deviation of the mean with the intention of expecting an interpretation that his mean stands a 68 per cent chance of being within this quantity of the "true value," a glance at Fig. 2.10 will convince the reader that, even with as many as 10 readings, the particular value for s can easily be so far from σ as to destroy the worth of the numerically exact statement that the observer was trying to make.

2.8 Practical Considerations

An understanding has now been reached of the nature of the result of repeating the measurement of a single quantity. To put this into practice in achieving optimum results from an experiment involves the balancing of the effort required to improve the intrinsic precision of the measurement itself against that of taking sufficient readings to improve the precision by a sort of brute force method. The use of the Expression (2.7) will enable a decision to be made regarding the number of duplicated readings required to achieve a certain specified precision. Here it must be noted that, since the standard deviation of the mean involves $\sqrt{n}$, the precision of the experiment improves rather slowly. It is rather tedious to obtain an additional significant figure by taking a hundred times as many readings. Usually it is easier, and it is certainly more satisfactory, to get a travelling microscope from the store room than it is to measure lengths with a meter stick to tenths of a millimeter. In any case, the theoretical reasoning cannot be pushed too far, since no amount of repetition of a meter stick reading will give a precision of a micron (i.e., 10^{-6} m). The finest scale division always provides a fundamental lower limit to the uncertainty.

The principal benefit, therefore, of repeated readings is not so much the improvement of the accuracy of the measurement as the provision of an estimate of the precision. Caution must be exercised in interpreting such estimates, however, since, as we have seen, the reliability with small numbers of readings is poor. It is useful to keep in mind the figures for the limits giving 68 per cent and 95 per cent confidence for 10 measurements. These limits of $\sigma/4$ and $\sigma/2$ will serve as a rough mental guide but, if a definite precision in σ value is required by the experimental conditions, the figures in Table 2 should be used to determine the numbers of observations required in the sample. It must be stressed again, however, that regardless of what is actually done in the experiment, the number of readings must be stated, in quoting the answer, to permit assessment of the significance of the result.

Of course, it may happen that duplication of the measurement does *not* give rise to differing answers. This is the case where the statistical fluctuations are sufficiently small to permit the distribution curve to shrink into the space between the finest scale divisions. In this case the result of the repetition will serve as evidence that the precision of measurement is limited only by the scale divisions. In all cases, however, the observer has to try repetition to discover which type of measurement he is dealing with, perturbation-limited or scale-limited.

In actual practice, then, the number of readings will be determined by a required precision of the mean or of the standard deviation. This number is normally so small that no conclusions about the actual distribution can be drawn. One therefore assumes that the set of observations is a sample from a Gaussian population. One quotes the mean as the best value and the standard deviation of the mean as the precision, and assumes that this last has the numerical significance appropriate to a Gaussian distribution.

2.9 Rejection of Readings

One last practical consideration of the distribution curve is concerned with outlying values. There is always the possibility of an observer making an actual mistake, perhaps in misreading a scale or in moving an instrument, accidentally and unnoticed, between setting and reading. There is always the temptation, therefore, to assign some such cause to a single reading which is well separated from an otherwise compact group of values. However, this is a dangerous temptation since the Gaussian curve does permit values remote from the central part of the curve, and also since the admission of the possibility of pruning of results makes it very difficult to know where to stop. The only answer can be in the judgement of the experimenter. This is not unreasonable since the measurement is his creation, and if his readers are going to give his conclusions any weight they are, in effect, expressing faith in his judgement. Many empirical rules for rejection of observations have been set up but they cannot replace personal judgement. It would be foolish to use a rule to reject one reading which was just outside the limit set by the rule, if there are other readings just inside it. There is also the possibility of extra information relating to the isolated reading which was noted at the time of making the reading, and which can help decide in favour of retention or rejection.

For a normal distribution curve the probability of obtaining a deviation greater than 2σ is 5 per cent (as we have seen before), greater than 3σ is $\frac{1}{3}$ per cent and greater than 4σ is no more than 6×10^{-5}. The criterion for rejection is still the responsibility of the observer, of course, but one can say, in general terms, that readings falling outside 3σ limits are more likely to be mistakes and so candidates for rejection. However, a problem arises because of our lack of information about the universe of readings and its constants X and σ. The better our knowledge of σ the more confident we can be that any far-out and isolated reading arises from a genuinely extraneous cause such as personal error in reading, malfunction of apparatus, etc. Thus, if we make 50 observations which cluster within 1 per cent of the central value and then obtain a reading which lies at a separation of 10 per cent, we would be fairly safe in suggesting that the circumstances had changed and that this last reading did not belong to the same universe as the previous 50. The necessity is to build up confidence in the main set of measurements before feeling justified in doing any rejecting. Thus, there is no justification for taking two readings and then rejecting a third on the basis of a 3σ criterion. Unless the situation is absolutely clear-cut it is by far the best to retain all the readings whether one likes them or not.

It is wise to remember also that many of the greatest discoveries in physics have taken the form of outlying observations.

2.10 Numerical Example

The considerations above of distribution curves and sampling are the keys to understanding the nature of measurement and it is of value to clarify our ideas by actual numerical examples. Let us pretend that the set of figures given in Table 1 on page 18 is a complete population for a particular experiment. Thus, by playing the role of Galileo, we shall be able to see just what luck the experimenter is going to have with his single sample. Since we are only pretending to have a complete set we cannot, of course, expect to get smooth distribution curves like Galileo's, but, even although rough, they will help to make the situation clear. We assume, therefore, that the distribution curve in Fig. 2.1 is that of the population, that the mean is the "true" value and that s for the set is σ. The mean can be found to be $X = 119$ and, using Equation (2.3), the standard deviation is found to be $\sigma = 12$. The distribution curve is replotted in Fig. 2.11 and the limits $\sigma = \pm 12$ from the mean are marked on it. Notice, first, how these 1σ limits divide

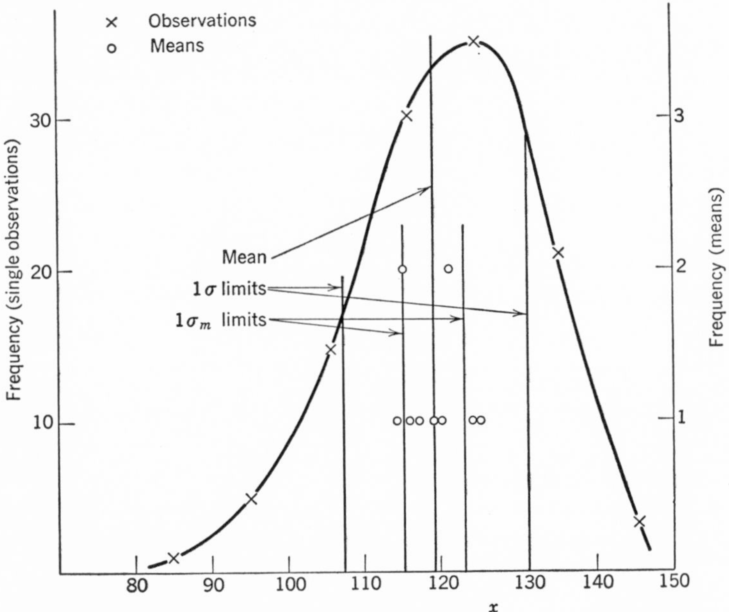

Fig. 2.11 Distribution of single readings and sample means.

the area under the curve to give approximately the 68 per cent fraction that we have been using, and so give an impression of the chances our experimenter would have with a single reading.

We now wish to perform Galileo's grouping process to investigate the sample fluctuations. Since the numbers in Table 1 were arranged in ascending order to aid in interpretation of the histogram, the numbers in the set were shuffled to give the impression of samples of random readings. Random grouping gives the following samples:

1	2	3	4	5	6	7	8	9	10	11
130	114	125	130	128	116	122	137	137	101	128
114	117	113	134	133	125	148	130	134	127	101
122	110	125	121	130	108	111	121	124	120	115
121	120	132	109	124	111	113	112	128	114	130
92	97	114	127	116	113	106	126	85	116	131
109	149	108	120	130	113	100	144	111	128	127
96	131	122	121	123	134	127	110	137	117	97
124	128	123	123	119	111	193	122	120	102	102
103	131	122	135	113	136	119	112	118	97	106
137	113	107	131	128	111	112	105	123	134	118

We first evaluate the means of the groups. They are:

1	2	3	4	5	6	7	8	9	10	11
114.8	121.0	119.1	125.1	124.4	117.8	116.1	120.5	121.6	115.6	115.5

We cannot plot much on a distribution curve with only 11 values, but by selecting a range of size unity, we can, as shown in Fig. 2.11, get an impression of how the means of the groups are clustered around the mean (119) of the population. This clustering shows us the degree of improvement in precision obtained by taking the mean of 10

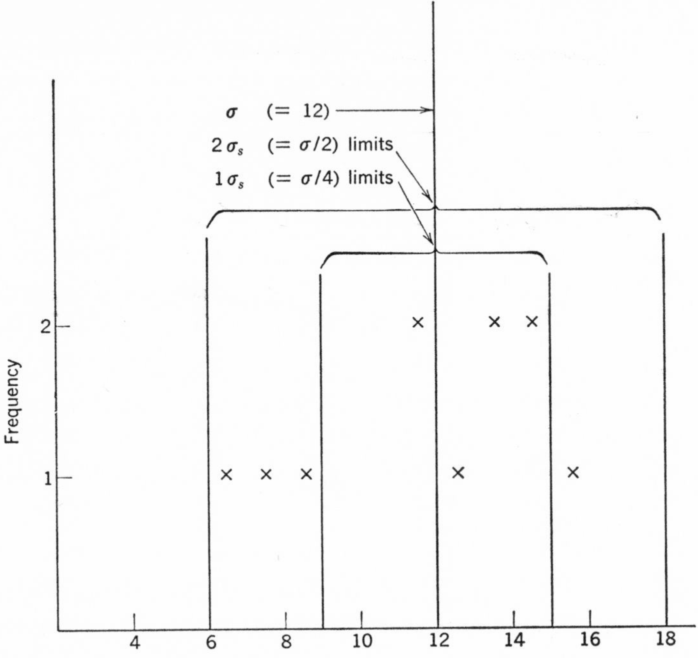

Fig. 2.12 Distribution of sample standard deviations.

values, instead of one single reading. We now evaluate the standard deviation of the mean from Equation (2.7). It is $12/\sqrt{10} = 3.8$. These limits of ± 4 are plotted about the mean on Fig. 2.11. Note how this value has the same kind of relationship to the distribution of means as σ had for the distribution curve of single values, and how it therefore serves as a measure of the improvement in precision which has resulted from the duplication of readings to form a sample.

To see what chance our observer, with his single sample, will have of getting a value of s reasonably close to σ, we shall evaluate s for each of the 11 samples using Equation (2.9). These standard deviations are:

1	2	3	4	5	6	7	8	9	10	11
14.7	14.4	8.2	7.9	6.6	11.3	14.0	11.4	15.4	12.4	13.2

Again, 11 values does not give a very good distribution curve, but by selecting a range of size unity, we can get an impression, as shown in Fig. 2.12, of how these sample standard deviations are distributed along the scale of s values and centered on 12, the value of the standard deviation of the population. Again we cannot expect too close numerical correspondence but we can believe, on looking at Fig. 2.12, that with more samples, 68 per cent of them would be contained within a range $\pm\sigma/4 = \pm 3$ about $\sigma = 12$, as suggested by Equation (2.8) and illustrated in Fig. 2.12, and that only 5 per cent would lie outside limits of $\pm\sigma/2 = \pm 6$.

As he looks at these illustrations, the reader should consider the situation of the observer with his single sample, and thus gain an impression of the chances involved in the result of the experiment. The most important lesson is the obviously high degree of sample fluctuation, even with as many as 10 readings in each sample. This should serve as the warning against the drawing of unjustifiable conclusions from sets of observations.

PROBLEMS

The following observations of angles (in minutes of arc) were made while measuring the thickness of a liquid helium film. Assume that the observations show random uncertainty and that they are a sample from a Gaussian universe.

34	35	45	40	46
38	47	36	38	34
33	36	43	43	37
38	32	38	40	33
38	40	48	39	32
36	40	40	36	34

1. Draw the histogram of the observations.

2. Identify the mode and the median.

3. Calculate the mean.

4. Calculate the best estimate of the universe standard deviation.

5. Calculate the standard deviation of the mean.

6. Calculate the standard deviation of the standard deviation.

7. Within which limits does a single reading have (a) a 68 per cent chance of falling, and (b) which limits give a 95 per cent chance?

8. Within which limits does the mean have (a) a 68 per cent chance, and (b) a 95 per cent chance of falling?

9. Within which limits does the sample standard deviation stand (a) a 68 per cent chance, and (b) a 95 per cent chance of falling?

10. Calculate a value for the constant h in the equation for the Gaussian error curve and for the probable error of the distribution.

11. If a single reading of 55 had been obtained in the set would you have decided in favor of accepting it or rejecting it?

12. Take two randomly chosen samples of five observations each from the main set of readings. Calculate their sample means and standard deviations to see how they compare with each other and with the more precise values obtained from the big sample.

13. If the experiment requires that the standard deviation of the mean should not exceed 1 per cent of the mean value, how many readings are required?

14. If the standard deviation of the universe distribution must be known within 5 per cent, how many readings are required?

3 The Propagation of Uncertainties

3.1 Absolute and Relative Uncertainty

The previous chapter has provided a means whereby an estimate of the uncertainty of measurement of a single quantity can be made. However, it is normally the case, in even the simplest experiments, that the final answer is to be computed in some way from values of several different quantities, each independently measured and subject, individually, to uncertainty. Thus, in the simple pendulum experiment, g is obtained as a function of T^2 and l, and the uncertainties in T and l will both contribute to an uncertainty in g. We are concerned with this resultant uncertainty in this chapter.

The word "uncertainty" will be used to signify outer limits of confidence within which we are "almost certain" (i.e., perhaps 99 per cent certain) that the measurement lies. This will commonly be valid only in simple measurements rounded off to the nearest scale division. When a measurement has been repeated often enough to give statistical significance to the result, one would obviously quote the standard deviation of the sample or the standard deviation of the mean, but we shall restrict the word "uncertainty" to the case of outer limits of confidence.

Let us suppose, therefore, that we have measured values of x and y (which we shall call x_0 and y_0) with either outer limits of uncertainty δx and δy or else standard deviations s_x and s_y. These uncertainties can be expressed in two forms, either of which can be the more useful, depending on circumstances. The uncertainty itself δx or δy will be referred to as an "absolute uncertainty." However, in judging the significance of this uncertainty in comparison with the actual value x_0, the ratio $\delta x/x_0$ (or s_x/x_0) is the more useful quantity. Obviously a $\frac{1}{2}$ mm uncertainty is far more significant when the measurement is 1 cm than when it is 1 m. This ratio will be referred to as the "relative uncertainty" or "precision" of the measurement. It will normally be expressed as a percentage.

3.2 Propagation of Uncertainty

The simplest case in which a result is computed from a measurement occurs when the result is a function of one variable only as, for example, the computation of the area of a circle from a measurement of a diameter.

Consider a computed result z to be a function of a variable x

$$z = f(x)$$

Here a measured value x_0 permits the required value z_0 to be calculated. However, the possibility that x can range from $x_0 + \delta x$ to $x_0 - \delta x$ means that there is a range about z_0 of possible values of z from $z_0 + \delta z$ to $z_0 - \delta z$. It is this value of δz which it is now desired to calculate. The situation is illustrated graphically in Fig. 3.1 in which it can be seen, for a given $f(x)$, how the measured value x_0 gives

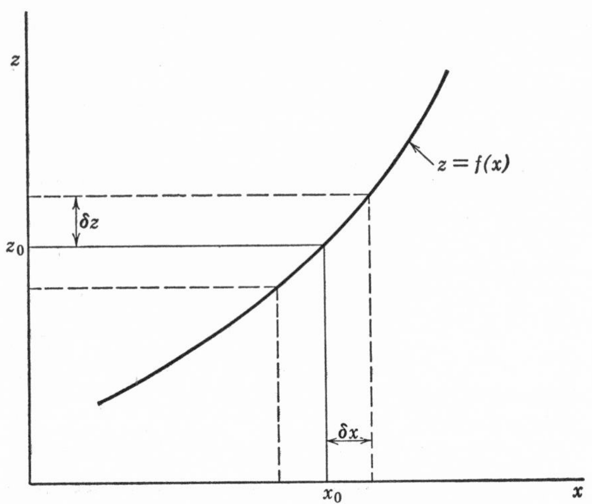

Fig. 3.1 Propagation of uncertainty from one variable to another.

rise to the computed result z_0, and how the range $\pm\delta x$ about x_0 gives a corresponding range $\pm\delta z$ about z_0.

Before proceeding to any general methods of evaluating δz, it is instructive to see how finite perturbations are propagated in simple functions. For example, consider the function

$$z = x^2$$

If x can range between $x_0 + \delta x$ and $x_0 - \delta x$ then z can range between $z_0 + \delta z$ and $z_0 - \delta z$ where

$$\begin{aligned} z_0 \pm \delta z &= (x_0 \pm \delta x)^2 \\ &= x_0^2 \pm 2x_0\,\delta x + (\delta x)^2 \end{aligned}$$

we can ignore $(\delta x)^2$, since δx is assumed to be small compared with x_0, and equate z_0 to x_0^2, giving us the value of δz as

$$\delta z = 2x_0\,\delta x$$

This can more conveniently be expressed in terms of the relative uncertainty $\delta z/z_0$ as

$$\delta z/z_0 = 2x_0\,\delta x/x_0^2 = 2\,\delta x/x_0$$

Thus, the relative uncertainty of the computed result is twice that of the initial measurement.

Although it is essential to bear in mind the nature of the propagation of uncertainty, as illustrated by this example with finite differences, a considerable simplification of the formulation can result from the use of the techniques of the differential calculus.

3.3 General Method for Uncertainty in Functions of a Single Variable

It will be noticed that these finite differences δz and δx are merely an expression of the derivative dz/dx. We can therefore obtain our value of δz by using standard techniques to obtain

$$\frac{dz}{dx} = f'(x)$$

and then writing

$$\delta z = f'(x)\,\delta x \tag{3.1}$$

This is a relatively simple procedure and will work in cases where the elementary finite difference approach would lead to algebraic complexity.

Thus, if
$$z = \frac{x}{x^2 + 1}$$

$$\frac{dz}{dx} = \frac{x^2 + 1 - x \cdot 2x}{(x^2 + 1)^2}$$

$$= \frac{1 - x^2}{(1 + x^2)^2}$$

$$\delta z = \frac{1 - x^2}{(1 + x^2)^2}\,\delta x$$

This would have been very awkward by any other approach. It gives δz generally as a function of x and δx, and the particular value desired would be obtained by setting $x = x_0$. Let us now use this technique to evaluate the uncertainty for some common functions.

(a) *Powers*

Consider
$$z = x^n$$
$$\frac{dz}{dx} = nx^{n-1}$$
$$\delta z = nx^{n-1}\,\delta x$$

The significance of this result becomes a little more obvious when expressed in terms of the relative uncertainty. Thus,

$$\frac{\delta z}{z} = n\,\frac{\delta x}{x}$$

This will hold for either powers or roots, so that the precision diminishes as a quantity is raised to powers or improves on taking roots. This is a situation which must be carefully watched in an experiment in which powers are involved. The higher the power, the greater is the initial precision that is needed.

(b) *Trigonometric Functions*

We shall do only one example since all the others can be treated in a similar fashion.

Consider
$$z = \sin x$$
$$\frac{dz}{dx} = \cos x$$
$$\delta z = \cos x\,\delta x$$

This is one case where the elementary method of inserting

$x_0 \pm \delta x$ shows the nature of the result more clearly. This substitution can be easily verified to give

$$\delta z = \cos x \sin \delta x$$

showing that the δx in the previous result is really $\sin \delta x$ in the limit. Only in the case of a very large uncertainty would this difference be significant, but it is best to understand the nature of the result. Clearly δx should be expressed in radian measure. The result will normally have straightforward application when dealing with apparatus such as the spectrometer.

(c) *Logarithmic and Exponential Functions*

Consider $$z = \log x$$

$$\frac{dz}{dx} = \frac{1}{x}$$

$$\delta z = \frac{1}{x}\,\delta x$$

and the relative uncertainty can be calculated as usual.

If $$z = e^x$$

$$\frac{dz}{dx} = e^x$$

$$\delta z = e^x\,\delta x$$

This is a rather more important case since the exponential function is one of common occurrence in physics and engineering. These functions can become very sensitive to the exponent when it takes values much over unity, and the uncertainty δz can be seen to have potentially large values. This will be familiar to anyone who has watched the current fluctuations in a thermionic diode which can result from quite small filament temperature variations.

As stated above, the method can be easily applied to any function not listed above by evaluating the appropriate derivative and using Equation (3.1).

3.4 Uncertainty in Functions of Two or More Variables

If the result is to be computed from two or more measured quantities, x and y, the uncertainty in the result can be regarded in two different ways. We can, first, be as pessimistic as possible and suppose that the actual deviations of x and y happen to combine additively in such a way that the value of z is driven as far as possible from the central value. We shall, in this way, calculate a δz which gives the extreme width of the range of possible z values. It is possible to argue against this that the probability is small of a number of uncertainties combining in magnitude and direction to give the worst possible result for z. This is true, and we shall deal later with the matter of the *probable* uncertainty in z. For the moment, however, let us calculate the δz which represents the widest range of possibility of z. This is certainly a safe, though pessimistic, approach since if δx, δy etc. represent limits within which we are "almost certain" the actual value lies, then this δz will give limits within which we are equally certain that the actual value of z lies.

The most instructive approach initially is to use the elementary substitution method, and we shall use this for the first two functions

(a) *Sum of Two or More Variables*

Consider

$$z = x + y$$

The uncertainty in z will be obtained from

$$z_0 \pm \delta z = x_0 \pm \delta x + y_0 \pm \delta y$$

and the maximum value of δz is given by choosing similar signs throughout. As might be expected, the uncertainty in the sum is just the sum of the individual uncertainties. This can be expressed in terms of relative uncertainties

$$\frac{\delta z}{z} = \frac{\delta x + \delta y}{x + y}$$

but no increased clarification is achieved.

(b) *Difference of Two Variables*

Consider

$$z = x - y$$

As in the case above, δz will be obtained from

$$z_0 \pm \delta z = (x_0 \pm \delta x) - (y_0 \pm \delta y)$$

Thus, we can obtain the maximum value of δz by choosing the negative sign for δy giving, once again,

$$\delta z = \delta x + \delta y$$

The significance of this is more clearly apparent if we consider the relative uncertainty given by

$$\frac{\delta z}{z} = \frac{\delta x + \delta y}{x - y}$$

This shows that, if x_0 and y_0 are close together, $x - y$ is small, and this relative uncertainty can rise to very large values. This is, at best, an unsatisfactory situation and it can become sufficiently bad to destroy the value of the measurement. It is a particularly dangerous condition since it can arise unnoticed. It is perfectly obvious that no one would attempt to measure the distance between two points a millimeter apart by measuring the distance of each from a third point a meter away, and then subtracting the two lengths. However, it can happen that a desired result is to be obtained by subtraction of two measurements made separately (two thermometers, clocks, etc.) and the character of the measurement as a difference may not be strikingly obvious. All measurements involving differences should be treated with the greatest caution. Clearly the way to avoid this difficulty is to measure the difference directly, rather than obtain it by subtraction between two measured quantities. For example if one has an apparatus within which two points are at potentials above ground of $V_1 = 1500$ v and $V_2 = 1510$ v respectively, and the required quantity is $V_2 - V_1$, only a very high quality voltmeter would permit the values of V_1 and V_2 to be measured to give $V_2 - V_1$ with even say 10 per cent. But an ordinary 10 v table voltmeter connected between the two points and measuring $V_2 - V_1$ directly will immediately give the answer with 2–3 per cent precision.

3.5 General Method for Uncertainty in Functions of Two or More Variables

These last two examples, treated by the elementary method, suggest that, once more, the differential calculus may offer a considerable simplification of the treatment. It is clear that if we have

$$z = f(x, y)$$

the appropriate quantity required in order to calculate δz is the total differential dz, given by

$$dz = \left(\frac{\partial f}{\partial x}\right) dx + \left(\frac{\partial f}{\partial y}\right) dy$$

We shall take this differential and treat it as a finite difference, δz, given, in terms of the uncertainties δx and δy, by

$$\delta z = \left(\frac{\partial f}{\partial x}\right) \delta x + \left(\frac{\partial f}{\partial y}\right) \delta y \tag{3.2}$$

where the derivatives $\partial f/\partial x$ and $\partial f/\partial y$ will normally be evaluated for the values x_0, y_0 at which δz is required. We shall find that the sign of $\partial f/\partial x$ or $\partial f/\partial y$ may be negative, in which case, using our pessimistic requirement for a maximum value of δz, we shall choose negative values for the appropriate δx or δy giving a wholly positive contribution to the sum.

(a) *Product of Two or More Variables*

Suppose

$$z = xy$$

Using Equation (3.2) we need

$$\frac{\partial z}{\partial x} = y \quad \text{and} \quad \frac{\partial z}{\partial y} = x$$

Thus, the value of δz is given by

$$\delta z = y\,\delta x + x\,\delta y$$

The significance of this is more clearly seen in the relative uncertainty

$$\frac{\delta z}{z} = \frac{\delta x}{x} + \frac{\delta y}{y}$$

i.e., when the result is a product of two variables, the relative uncertainty of the result is the sum of the relative uncertainties of the components.

The most general case of a compound function, and one very commonly found in physics, is the one in which an algebraic product has components raised to powers in the form

$$z = x^a y^b$$

where a and b may be positive or negative, integral or fractional powers. In this case the formulation is greatly simplified by taking logs of both sides before doing the differentiating.
Thus,

$$\log z = a \log x + b \log y$$

Therefore, differentiating implicity,

$$\frac{dz}{z} = a\frac{dx}{x} + b\frac{dy}{y}$$

As usual, we take the differentials to be finite differences, giving

$$\frac{\delta z}{z} = a\frac{\delta x}{x} + b\frac{\delta y}{y}$$

Note that this process gives the relative uncertainty directly. This is frequently convenient but, if the absolute uncertainty δz is required, it is simply evaluated by multiplying by the computed value z_0, which is normally available. This form of implicit differentiation is still the simplest even when z is itself raised to some power. For if the equation reads

$$z^2 = xy$$

it is unnecessary to rewrite it

$$z = x^{1/2}y^{1/2}$$

and work from there because, by taking logs,

$$2 \log z = \log x + \log y$$

i.e.,

$$2 \frac{\delta z}{z} = \frac{\delta x}{x} + \frac{\delta y}{y}$$

giving $\delta z/z$ as required.

(b) *Quotients*

These come under the heading of the previous section, which permits negative values, and we repeat that the maximum value of δz will be obtained by neglecting the negative sign in the differential.

If a function other than those already listed is encountered, some kind of a differentiation will usually be found to work. It is frequently a convenience to differentiate an equation implicitly, thus simplifying the working by avoiding the necessity for calculating the unknown explicitly as a function of the other variables. For example, consider the lens equation

$$\frac{1}{f} = \frac{1}{s} + \frac{1}{s'}$$

where f is a function of the measured quantities s and s'. We can differentiate the equation implicitly to obtain

$$-\frac{df}{f^2} = -\frac{ds}{s^2} - \frac{ds'}{s'^2}$$

It is now possible to calculate df or df/f directly and more easily than would have been the case by writing f explicitly as a function of s and s'. Thus, a formula may be prepared for the uncertainty into which all the unknowns can be inserted directly. Make sure that the appropriate signs are used so that the contributions to the resultant uncertainty all add positively to give the outer limits of possibility for the answer.

If the function is too big and complicated to work out a value of δz in general, one can always take the measured values x_0, y_0 and work out z_0. Then if one evaluates the result by substituting the actual numerical values of $x_0 + \delta x$, $y_0 + \delta y$ (or $y_0 - \delta y$ if appropriate) to give one of the outer values of z and then repeating the other way, the limits on z have been determined and δz obtained.

3.6 Compensating Errors

A special situation can arise when compound variables are involved. Consider, for example, the well-known relation for the angle of minimum deviation D in a prism of refractive index μ and vertical angle A

$$\mu = \frac{\sin \frac{1}{2}(A + D)}{\sin \frac{1}{2} A}$$

If A and D are measured variables with uncertainties δA and δD, the quantity μ will be the required answer, with an uncertainty $\delta\mu$. It would be fallacious, however, to calculate the uncertainty in $A + D$, then in $\sin \frac{1}{2}(A + D)$, and combine it with the uncertainty in $\sin \frac{1}{2} A$, treating the function as a quotient of two variables. This can be seen by thinking of the effect on μ of an increase in A. Both $\sin \frac{1}{2}(A + D)$ *and* $\sin \frac{1}{2} A$ increase, and the change in μ is not correspondingly large. The fallacy is in the application of the particular methods of the previous sections to variables which are not independent (e.g., $A + D$ and A). The cure is either to reduce the equation to a form in which the variables are all independent, or else to go back to first principles and use the equation of Sec. 3.5 directly.

Cases which involve compensation of errors should be watched carefully since they can, if treated incorrectly, give rise to large errors in uncertainty calculations.

3.7 Standard Deviation of Computed Values: General Methods

As has been frequently stressed, this last section has been concerned with outer limits of possibility for the computed value z. We have already suggested that this represents an unrealistically pessimistic approach and that the more useful quantity would be a *probable* value for δz, provided we can attach a numerical meaning to "probable." The limits given by this quantity will be smaller than $\pm\delta z$, but we have the hope of an actual numerical significance for them. Such statistical validity will be possible only if the uncertainties in x and y have such validity, and we shall, therefore, assume that the measurements have been sufficiently numerous to justify a calculation of the standard deviation of the x values s_x, and correspondingly, of s_y. We then hope to be able to calculate an s_z.

However, we must first inquire what we mean by s_z. We assume that the measurement has taken the form of pairs of observations x, y (for example, the current through and the potential across a resistor, which have been measured with the aim of obtaining the resistance) obtained by repetition under the same conditions. Each pair will define a value of z and, if the repetition had yielded n pairs, we shall have a set of n values of z showing statistical fluctuations. The quantity we require, s_z, is the standard deviation of this set of z values. Now these individual z values may never be calculated, because one would calculate the means $\bar{x}$ and $\bar{y}$ and obtain $\bar{z}$ directly using the assumption (valid if s_x, s_y and s_z are small compared, respectively, with x, y, and z) that

$$\bar{z} = f(\bar{x}, \bar{y})$$

Nevertheless, that is the significance of the s_z we are about to calculate.

If we assume that the universes of the x, y, and z values have a Gaussian distribution, the quantity σ_z (of which we are about to calculate the best estimate in terms of s_z) will have the usual significance that any z value will stand a 68 per cent chance of falling within $\pm\sigma_z$ of the true value.

As before, let

$$z = f(x, y)$$

and consider perturbations δx, δy which lead to a perturbation δz given by

$$\delta z = \left(\frac{\partial z}{\partial x}\right) \delta x + \left(\frac{\partial z}{\partial y}\right) \delta y$$

This perturbation can be used to calculate a standard deviation for the n different z values since

$$s_z = \sqrt{\Sigma\, (\delta z)^2/n}$$

Thus

$$\begin{aligned} s_z^2 &= \frac{1}{n} \Sigma \left[\left(\frac{\partial z}{\partial x}\right) \delta x + \left(\frac{\partial z}{\partial y}\right) \delta y\right]^2 \\ &= \frac{1}{n} \Sigma \left[\left(\frac{\partial z}{\partial x}\right)^2 (\delta x)^2 + \left(\frac{\partial z}{\partial y}\right)^2 (\delta y)^2 \right. \\ &\qquad \left. + 2\left(\frac{\partial z}{\partial x}\right)\left(\frac{\partial z}{\partial y}\right) \delta x\, \delta y\right] \end{aligned}$$

$$= \left(\frac{\partial z}{\partial x}\right)^2 \frac{\Sigma\,(\delta x)^2}{n} + \left(\frac{\partial z}{\partial y}\right)^2 \frac{\Sigma\,(\delta y)^2}{n}$$

$$+ \frac{2}{n}\left(\frac{\partial z}{\partial x}\right)\left(\frac{\partial z}{\partial y}\right) \delta x\,\delta y$$

But $$\frac{\Sigma\,(\delta x)^2}{n} = s_x^2 \quad \text{and} \quad \frac{\Sigma\,(\delta y)^2}{n} = s_y^2$$

and, since δx, δy may be considered for the present purpose to be independent perturbations,

$$\Sigma\,\delta x\,\delta y = 0$$

Thus, finally

$$s_z = \sqrt{(\partial z/\partial x)^2 s_x^2 + (\partial z/\partial y)^2 s_y^2} \tag{3.3}$$

If z is a function of more than two variables the equation is extended by adding similar terms.

Thus, if the components of a calculation have standard deviations of some degree of reliability, a value can be found for the probable uncertainty of the answer where "probable" has a real numerical significance.

The calculation has been carried out in terms of the variance or standard deviation of the x and y distributions. However, in actual practice the quantities we want are the best estimates of σ_x, σ_y, etc., and so we would use the modified value with denominator $n - 1$ in accordance with Equation (2.9). The result would then be a best estimate for σ_z. The standard deviation of the mean for z can then be calculated by direct use of Equation (2.7) and this will give the limits within which the *mean* value of z, $\bar{z}$, stands a 68 per cent chance of falling.

Note that most actual experiments do not accord with the assumptions of the development just given. If we are measuring the flow rate of water through a pipe, we shall measure the flow rate, pipe radius and pipe length independently and each one with a number of readings dictated by the intrinsic precision of the measurement. We cannot, therefore, use Equation (3.3) directly, since the various s's are different types of quantity. The solution is to calculate the standard deviation of the mean for each of the elementary quantities first. If these are used in Equation (3.3), the result of the calculation will be immediately a standard deviation of the mean for z.

3.8 Standard Deviation of Computed Values: Special Cases

Let us now apply Equation (3.3) to a few common examples. In all the following cases the various s's are all assumed to be best estimates of the appropriate universe value σ.

(a) *Sum of Two Variables*

$$z = x + y$$

hence
$$\frac{\partial z}{\partial x} = 1, \quad \frac{\partial z}{\partial y} = 1$$

and
$$s_z = \sqrt{s_x^2 + s_y^2}$$

Note that this result affords a justification for Equation (2.7) on page 33. The mean value for the sample, $\Sigma\, x_i/n$, is just a function such as $z = x + y$, where x and y happen to be independent measurements of the *same* quantity. Thus if

$$z = \frac{1}{n}(x_1 + x_2 + x_3 + \cdots)$$

$$\frac{\partial z}{\partial x_1} = \frac{1}{n}, \quad \frac{\partial z}{\partial x_2} = \frac{1}{n}, \quad \text{etc.}$$

and $$s_z = \sqrt{\left(\frac{1}{n}\right)^2 s_x^2 + \left(\frac{1}{n}\right)^2 s_x^2 + \cdots}$$

$$= \sqrt{ns_x^2/n^2} = \frac{s_x}{\sqrt{n}}$$

(b) *Difference of Two Variables*

$$z = x - y$$

Here $$\frac{\partial z}{\partial x} = 1, \quad \frac{\partial z}{\partial y} = -1$$

but again $$s_z = \sqrt{s_x^2 + s_y^2}$$

As dealt with in Sec. 3.4 on page 56, the previous considerations regarding measurements of differences are still valid.

(c) *Product of Two Variables*

$$z = xy$$

hence $$\frac{\partial z}{\partial x} = y, \quad \frac{\partial z}{\partial y} = x$$

thus $$s_z = \sqrt{y^2 s_x^2 + x^2 s_y^2}$$

and the specific value for s_z at the particular values x_0, y_0 of x and y would be obtained by substituting x_0 and y_0 in the equations.

Just as in the previously treated case of products, the equation is more clearly expressed in terms of relative values of s i.e. s_z/z. We obtain

$$\frac{s_z}{z} = \sqrt{s_x^2/x^2 + s_y^2/y^2}$$

(d) *Variables Raised to Powers*

$$z = x^a$$

$$\frac{\partial z}{\partial x} = ax^{a-1}$$

$$s_z = \sqrt{a^2 x^{2(a-1)} s_x^2}$$

Again this is more instructive when expressed in terms of the relative value

$$\frac{s_z}{z} = \sqrt{a^2 \frac{s_x^2}{x^2}}$$

$$= a \frac{s_x}{x}$$

(e) *The General Case of Powers and Products*

$$z = x^a y^b$$

Obviously the results of (c) and (d) can be extended to give the result

$$\frac{s_z}{z} = \sqrt{\left(a \frac{s_x}{x}\right)^2 + \left(b \frac{s_y}{y}\right)^2}$$

In this result note that the presence of negative indices in the original function is unimportant, since they occur only squared in the expression for s_z.

If a function other than those listed above is encountered, the use of Equation (3.3) will yield the required result. It can be seen that, for the case of a function of a single variable, $z = f(x)$, Equation (3.3) reduces to the same form as that for uncertainties, Equation (3.1). The result is, therefore, the same for standard deviations as it was for uncertainties in the case of the trigonometric, exponential and logarithmic functions treated in Sec. 3.3.

Note that, although we listed in Secs. 3.2 to 3.5 a number of different approaches to the problem of outside limits to uncertainty, the standard deviation of z is a uniquely defined quantity and there is no alternative to the use of Equation (3.3).

3.9 Combination of Different Types of Uncertainty

Unfortunately for the mathematical elegance of the development, it very frequently occurs that the uncertainty in a computed result is required when the component quantities have different types of uncertainty. Thus we may require the uncertainty in

$$z = f(x, y)$$

where x is a quantity to which have been assigned outer limits $\pm\delta x$ within which we are "almost certain" that the actual value lies and y is a quantity whose uncertainty is statistical in nature, and for which a sample standard deviation s_y and a standard deviation of the mean $s_y/\sqrt{n}$ have been calculated. We require the uncertainty in z. The problem is that the uncertainty in z is a difficult thing even to define. We are trying to combine two quantities which have, in effect, completely different distribution curves. One is the standard Gaussian function but the other is a rectangle, bounded by the outer limits of uncertainty, and flat on top because the actual value of the unknown x is equally likely to be anywhere between the outer limits $x_0 \pm \delta x$. Any general method of solving this problem is likely to be far too complex for general use, but particular solutions can be found following a method suggested by Dr. T. M. Brown.

In the calculation for z one uses the sample mean $\bar{y}$ for the y value. This has the significance that it stands approximately a $\frac{2}{3}$ chance of coming within $\pm s_y/\sqrt{n}$ of the true value. Let us therefore calculate limits for x which, similarly, give a $\frac{2}{3}$ probability of enclosing the true value. Since the probability distribution for x is rectangular, $\frac{2}{3}$ of the area under the distribution curve is enclosed by limits which are separated by a distance equal to $\frac{2}{3}$ of the total range of possibility, i.e., $\frac{2}{3}$ of $2\ \delta x$. The limits for $\frac{2}{3}$ probability are therefore $\frac{4}{3}\ \delta x$ or $\pm\frac{2}{3}\ \delta x$.

This quantity, $\frac{2}{3}\ \delta x$, is therefore one which can be compared with $s_y/\sqrt{n}$, since both quantities correspond to $\frac{2}{3}$ probability. Equation (3.3) can now be used, inserting $\frac{2}{3}\ \delta x$ for the value of the standard deviation of the mean for x and $s_y/\sqrt{n}$ for the y function. This will yield a quantity for the uncertainty in z which has a $\frac{2}{3}$ probability attached to it, and will serve instead of an s_z. Note, however, that it would not be true to say that 95 per cent probability would be represented by limits twice as widely spaced as those calculated for $\frac{2}{3}$ probability. The limits for 95 per cent probability would have to be calculated separately using the method above.

3.10 Application of Results

Two approaches to the problem of calculating the uncertainty of a computed value have been used in this chapter. In the first the pessimistic calculation was made of the outer limits of possibility for the answer. The use of such a calculation is restricted to cases in which there does not exist sufficient precision of measurement to justify the calculation of a standard deviation. This would be the case where the scale is not divided with sufficient fineness to permit the statistical fluctuations to be observed. The use of the outer limits would also be appropriate in the preliminary analysis of an experiment, as described later, to serve as a guide to the conduct of the experiment. The use of the second approach is limited, as stated above, to cases of genuine statistical significance. These will usually be encountered in the evaluation of an experiment on completion of the collection of observations. Here the emphasis will be on the precision actually achieved, and a quantity with numerical significance will be sought.

PROBLEMS

1. A meter stick (read to the nearest mm) is used to measure a length of 12 cm. What is the absolute uncertainty? What is the relative uncertainty?

2. A travelling microscope can be read to 0.1 mm. What is the precision of the measurement of a distance of 1 cm?

3. What is the smallest distance which can be measured using a meter stick (read to mm) so that the uncertainty shall not exceed (a) 1 per cent, (b) 5 per cent?

4. A distance of 2 cm must be measured to 1 per cent. (a) Would a meter stick be suitable? (b) Would a travelling microscope (read to $\frac{1}{10}$ mm)?

5. A barometer reading normal atmospheric pressure can be read to 0.1 mm. What is the precision?

6. An ammeter reading 0–5 amp is graduated in 0.1 amp. Assuming that it is read to the nearest scale division, what is the precision of measurement (a) at full scale? (b) at 1 amp?

7. A stop watch is graduated in $\frac{1}{5}$ sec. What is the minimum time interval which can be measured with a precision of (a) 5 per cent, (b) 0.1 per cent?

8. A wrist watch gains 1 min/day. What is the precision with which it can be used to time an interval of 1 hr?

9. It is stated that today is 5.4° warmer than yesterday. Both measurements were made on the same thermometer read to 0.2°. What is the precision of the statement?

10. In measuring a resistance the voltage is read as 5.4 v and the current as 1.3 amp (both read to the first decimal place). What is the absolute uncertainty of the resistance value?

11. A density measurement gives the following figures: mass, 24.32 g $\pm$ 0.005; volume, 10.2 $\pm$ .05 cc. What is the absolute uncertainty in the density?

12. A simple pendulum experiment to measure g using $T = 2\pi \sqrt{l/g}$ gave T to 2 per cent and l to 1.5 per cent. What is the precision of the g value?

13. Young's modulus Y for a material can be found from the deflection of a loaded beam using the equation

$$d = \frac{4Wl^3}{Yab^3}$$

$$\begin{aligned} d = \text{deflection} &= 14.2 \pm 0.1 \text{ cm} \\ W = \text{load} &= 500 \text{ g (exact)} \\ a = \text{beam width} &= 2.1 \text{ cm} \pm 0.05 \\ b = \text{beam thickness} &= 0.3 \text{ cm} \pm 0.05 \\ l = \text{beam length} &= 45.1 \text{ cm} \pm 0.1 \end{aligned}$$

What is the absolute uncertainty in the measurement of Y?

14. The focal length of a thin lens is measured using the equation

$$\frac{1}{s} + \frac{1}{s'} = \frac{1}{f}$$

s is found to be 24.3 $\pm$ 0.05 cm, s' 17.4 $\pm$ 0.05 m. What is the precision of the f measurement?

15. Using a diffraction grating for which

$$d \sin \theta = \lambda$$

angles θ are measured to 1 min of arc. A wavelength λ is determined from a θ measurement of 15° 35′. What is the relative uncertainty of the λ measurement? (d can be considered to be precise).

16. A-c measurements at angular frequency ω are made on a series circuit of a resistor and inductor. The impedance Z is given by

$$Z = \sqrt{R^2 + \omega^2 L^2}$$

The resistance R is known to be 50 ohms with a precision of 5 per cent, L is known to be 2 henry with a precision of 10 per cent and ω is exactly $2\pi \times 60$. What is the absolute uncertainty in the measurement of Z?

17. Ice was added to water in a measurement of the heat of fusion, H, giving the equation

$$m_i H + m_i T_2 = m_w (T_1 - T_2)$$

m_i = mass of ice = 14.2 ± 0.1 g
T_1 = initial temperature of water = $25.4° \pm 0.1$
T_2 = final temperature of water = $7.8° \pm 0.1$
m_w = mass of water = 72.3 ± 0.1 g

What is the absolute uncertainty in the measured heat of fusion?

18. The coefficient of linear expansion α of a solid is to be measured using the equation

$$l = l_0 (1 + \alpha \, \Delta T)$$

The length l_0 is about 50 cm and the expansion $l - l_0$ can be measured to 5/100 mm. Knowing that α is about 2×10^{-5} and neglecting the uncertainties in measuring l_0 and ΔT, calculate the minimum temperature range ΔT which will permit α to be measured to 10 per cent.

19. A 1 m slide wire is used as two arms of a Wheatstone bridge so that an unknown resistance R is calculated from

$$R = \frac{l_1}{l_2} R_s$$

R_s, the standard resistor is 10.0 ± 0.05 ohms, l_1 is 72.3 ± 0.05 and $l_1 + l_2 = 100$ exactly. What is the absolute uncertainty in R?

20. The heat capacity, S, of a liquid is measured by a continuous flow calorimeter for which

$$VI = JSQ(T_2 - T_1)$$

V and I can be read to 2 per cent each, J is known exactly, Q can be measured to $\frac{1}{2}$ per cent and the thermometers for T_1 and T_2 can be read to $\pm 0.1°$. What is the minimum value of $T_2 - T_1$ which will allow S to be measured to 10 per cent?

21. The decay constant of a ballistic galvanometer can be obtained from the equation

$$\theta_2 = \theta_1 e^{-kT}$$

where θ_1 and θ_2 are successive deflections (on the same side) in the damped oscillation and T is the period of vibration. T was measured to be 5.4 sec $\pm$ 0.1 and θ_1 and θ_2 were observed to be 24.1 and 16.5, each measurement being uncertain to ± 0.2. What is the uncertainty in the k measurement?

22. The resistance R of a parallel network of two resistors R_1 and R_2 is given by

$$R = \frac{R_1 R_2}{R_1 + R_2}$$

where $\quad R_1 = 5.4 \pm 0.1, \quad R_2 = 1.4 \pm 0.05$

In the form of the equation given here this is a compensating error case. Work it out the wrong way (i.e., finding the uncertainty in $R_1 R_2$ and in $R_1 + R_2$ and combining) and the correct way so as to see the difference.

23. An experiment to measure the refractive index μ of a prism uses the relation

$$\mu = \frac{\sin \frac{1}{2}(A + D)}{\sin \frac{1}{2}A}$$

The angle A is measured to be $60° \pm 2'$; the angle D is $23° 14' \pm 2'$. What is the uncertainty in μ?

24. Repeated measurements of the diameter of a wire of circular cross section gave a mean of 0.41 mm with a sample standard deviation of 0.07. What is the sample standard deviation of the resulting calculation of the cross-sectional area?

25. The wavelengths of the two yellow lines in the sodium spectrum are measured to be 5891.1 Å with a standard deviation of 1.5 Å and 5896.8 with a standard deviation of 1.5 Å. What is the standard deviation for the wavelength difference between the two lines?

26. A potentiometer with a 1 m wire is being used to measure the emf of a cell in terms of that of a standard which is known exactly as 1.0183 v. Only a rather insensitive nullpoint galvanometer was available and 20 determinations of the balance point gave a mean of 68.3 cm with a sample standard deviation of 1.4 cm. Using

$$V_{\text{unknown}} = \frac{l_1}{l_2} V_{\text{standard}}$$

calculate the standard deviation of the mean for V_{unknown}.

27. A simple pendulum is used to measure g using

$$T = 2\pi \sqrt{l/g}$$

20 measurements of T give a mean of 1.82 sec and a sample standard deviation of 0.06. Ten measurements of l give a mean of 82.3 cm and a sample standard deviation of 1.4. What is the standard deviation of the mean for g?

28. The coefficient of viscosity of water is being measured by Poiseuille's equation

$$Q = \frac{P\pi a^4}{8\eta l}$$

P has been measured to be definitely between 17 and 18 cm of water pressure, Q to be within ± 1 of 204 cc/min, l was measured with a meter stick (read to mm) to be 32.1 cm, a was difficult to measure and was observed 10 times yielding a mean of 1.2 mm and a sample standard deviation of 0.2 mm. Within which limits are we 68 per cent certain that the value of η lies?

4 The Nature of Experimenting

4.1 Nature of Scientific Theory

An experiment can be defined as any planned observational process by which man increases his experience of the external world. This definition is really so broad as to be useless, but it does serve to emphasize the enormous range of human activity which comes under the heading of experimenting. Because of this range, the advice which could be given to help a student is likely to be so generalized and vague as to be of very little practical assistance. Even so, the nature of the experimental process in principle, and its role in establishing the nature of human knowledge should be part of the education of everyone who is likely to be faced with the problem of gathering observational information of any type. For the present purpose we shall restrict ourselves to a few general remarks and then consider especially the principles of experimenting in the physics laboratory.

The general course by which a science grows is normally an alternation of experimenting and thinking. The sequence is usually observation-hypothesis-experiment. The first step in the sequence is often the chance observation of a new phenomenon. Since this is automatically outside the range of previous human experience, the observation is naturally followed by speculation regarding the nature of the phenomenon. This speculation will probably result in some genius introducing a new idea which has been stimulated by the observation. For example, the observation might be the mode of scattering of α particles from a gold foil, and the resultant proposal is that of Rutherford's nuclear atom. Note that our theoretician could, in principle, have thought of this without knowledge of the experimental discovery. The experiment thus acts only as a trigger on the imagination of the scientist but, because of the vast range of speculative ideas which the theoretical scientist could have produced, the observation is an absolutely essential guide. It may be that the axiom which is accepted is not a completely new idea but is nothing more than an idealization of observed behavior e.g., Ohm's law as a basis for circuit theory, or Hooke's law in the theory of elastic media. This axiomatic foundation for the development is frequently called a "model."

Such exercise of the imagination by our theoretical scientist is perhaps amusing but not really profitable until he can develop the idea to the point when he can make a prediction about some aspect of the phenomenon which has not yet been treated experimentally. This development gives the scientist's idea the status of a theory, and the predictions made will immediately suggest experiments. The requirement to subject theoretical predictions to experimental test can be used to define the "scientific method." The results of these experiments decide whether the original speculation was fruitful or not. If it was a good idea, the predictions of the theory will conform with the experimental observations. This does not mean, of course, that the theory is "right," since the only evidence we have in its favor is correspondence with experiment within a certain degree of instrumental precision. It would be a bold scientist who ever claimed that a theory represented absolute truth about nature. If the so-called laws of physics of the present day seem unshakable to the reader, he should reflect a little on the strength of the convictions held by the propounders of the Ptolemaic system of planetary orbits or of phlogiston. One can, however, prove ideas to be *wrong* more or less easily. The postulate that nights are dark because the sun is swallowed by a dragon at the end of each day is a very satisfactory one, for which every evening provides fresh confirmation. It is satisfactory, however, only so long as the observer stays in one place. An expedition to the opposite side of the earth could easily relay radio signals back home which would transmit evidence that the sun had not been swallowed but had merely gone around the corner. In general, then, one can prove ideas about nature to be wrong but one cannot prove them to be right.

Note that this concept of rightness or wrongness applies only when we wish to make a statement about the natural world. Once the basis for a theory has been postulated (such as Euclidean geometry or Newtonian mechanics) there is no question about the "truth" of the deductions, where one defines "truth" as "correctly derived from the axioms." These deductions, however, are statements about the *theory*, not about the natural world. This philosophical question arises only at the frontiers of knowledge where questions of the nature of the universe are considered. In everyday work in the laboratory one is not so much concerned with "rightness" or "wrongness." It is better to talk of "appropriateness," because even supplanted theories are used extensively, normally because they tend to be simpler than their replacements. The introduction of general relativity by Einstein does not destroy the usefulness of the Newtonian inverse square force "law," and the introduction of wave mechanics does not prevent one from sometimes thinking of electrons as little charged billiard balls. One uses theories more on a basis of convenience than anything else, provided they give satisfactory correspondence with experiment at the level of precision being considered.

It is frequently found, when subjecting a theory to experimental investigation, that one does not find complete correspondence, or lack of it, between prediction and observation. More commonly it is found that the original speculation is partly satisfactory. If the idea is too elementary (as it is almost bound to be when first propounded) then the experiment will probably show correspondence with the predictions within a limited range only. This partial discrepancy then constitutes another observation which starts the cycle all over again.

The structure of scientific thought is therefore a complex of deductions from a set of observation-stimulated axioms. Sometimes the axiomatic nature of the science is clear, as in the case of Euclidean geometry, but, even in the most empirical and observational sciences it is still there. One discusses the resistance of a metal wire in terms of Ohm's law or the motion of electrons in a cathode ray tube using the charged billiard ball model. It is important that this axiomatic structure be recognized, and remembered, so that the significance of any scientific or technological statement or conclusion may be judged. The axiomatic foundation provides the background of thought to which all experimental processes are referred.

4.2 Types of Experiment

As has been stated, the range of activity covered by the term experimenting is so broad that a detailed description is almost useless. There are, however, some considerations which enter into every type of work, and which help to influence the conduct of the work. These considerations, listed below, will enter an experimental situation in varying degree and any practical problem will be constructed out of them.

The three main factors involved in determining the type of experiment are:—completeness of background material, degree of control over the subject material, and degree of statistical fluctuation. The influences of these will be considered in turn.

(a) *Degree of Reference to Background*

In simple terms this is nothing more than the degree of familiarity with the subject of the experiment. It obviously can vary widely between extreme limits of complete familiarity and complete strangeness. It can be quite close, as in the case of a measurement of a quantity using a theory, concept or principle which is well established. An example of such a situation would be the measurement of the electrical resistivity of a copper wire using a Wheatstone bridge. In this case every aspect of the measurement is well-tried, familiar and, within certain limits of precision, the answer is unequivocal. No difficulties in interpretation arise because everyone knows what the experimenter is talking about.

However, at the other end of the scale the degree of reference to background material may be very limited since, in the case of a newly discovered phenomenon there is no background material which is directly relevant. The study of this type of situation constitutes exploratory type research, and this is very difficult work just because of the lack of guidance. The aim of the work is, in this case, to gather as much information as possible (usually as quickly as possible because of human impatience) covering as wide a range of approach as possible. This will limit the range of speculation about the phenomenon and increase the probability of someone making a lucky guess. Prior to this lucky guess the interpretation of the measurements is very difficult in view of the lack of guidance towards such an interpretation. The only course open to the experimenter is to try to find some function which fits the observation and which he hopes will act as a guide to theoretical thought. In this he may be optimistic since many empirically established functions have turned out later to have no relation whatsoever to the theoretical functions later developed.

The point to be noted is that, although the familiarity of the concepts of an experiment range along the whole scale as described above, some degree of reference to background ideas is always found. This may be obvious half way along the familiarity scale where one of the points of an experiment may be the establishment of the validity of the concepts on which the experiment is based. One would not quote a measured value for the half life of a radioactive isotope without verifying an exponential law for the decay of its activity. However, the degree of reference to background material may be forgotten at the two ends of the familiarity scale, and for a different reason each time. At the familiar end one tends, just because of the familiarity, to forget the axiomatic foundation of the measurement. The observer with his Wheatstone bridge may forget that he is automatically invoking Ohm's law. Even such a seemingly clear-cut measurement as the distance between two buildings on the University campus involves an interpretation of the theodolite readings in terms of the axioms of Euclidean geometry. At the other end of the familiarity scale the experimenter has to be especially careful with the interpretation of his newly discovered and still obscure phenomenon. He will automatically use his already existing ideas in considering it, just because, from the very nature of the situation, he has nothing else available. There is, however, no guarantee that the earlier ideas have any validity for the new phenomenon. Even such a startling discovery as that of superconductivity in 1911 was, for long, interpreted in terms of the established concepts of conductivity. The inappropriateness of such an interpretation will be revealed only by the development of a theory based on it and subsequent comparison with experiment.

To summarize, we repeat that, even though the degree of reference to background in an experiment may vary, it is always there. The concept of the impersonal, objective measurement of the "hard facts of science" is not appropriate. All measurements and experiments are conceived and interpreted in terms of the current modes of thought.

(b) *Degree of Control*

The degree of control which the experimenter has over his experimental material varies enormously. The production staff of a large-scale industrial process which is not functioning correctly may be able to do little more than watch it in the hope of making a guess at the fault. On the other hand, a metal specimen in a solid state physics laboratory may have every known physical property wrung out of it. In between, there is every gradation of control, and normally the degree of control is fixed by the circumstances of the experiment. The student must learn to accept the available extent of his influence, or lack of it, over the experiment and surmount this difficulty in obtaining his result.

As one might expect, the difficulties arise under conditions where control over the actual conduct of the experiment is poor. Here the whole secret of success is in preliminary planning because, once the experiment is under way, the experimenter may be able to do little more than watch its progress helplessly. The planning of the experiment will determine if it is going to be successful, and, with poor planning, the results may not be merely uninformative but actually may be misleading.

The physics laboratory is misleading to the student in this respect because, if the student makes a mess of an experiment, the opportunity is commonly available to retrieve any missing readings or even repeat the whole experiment. Under actual working conditions, however, this may not be so easy. A scientist who has been given the job of measuring the expansion of nuclear reactor fuel rods under neutron irradiation would be most unpopular if, after wasting many thousands of dollars worth of neutrons in the reactor, he found he had forgotten to measure the initial size of his rods. Experiment planning will be considered in more detail in the next chapter and it will suffice for the moment to say that the student will find it well worth while acquiring, at the earliest possible stage, the habit of detailed planning of an experiment.

The degree of control available during the actual conduct of the experiment is low in large-scale industrial processes, as has already been suggested. It is also low in principle in those experiments of an essentially observational type, such as are common in sociological work (e.g., What is the most popular brand of toothpaste in North America? For the average observer this is not a controllable variable.) In cases like this the problem reduces, almost invariably, to one of sampling, and once again the essence of work is planning, because it is the only activity open to the experimenter. The fallibility of such procedures consequent on poor planning is well known. Such fallibility can arise either as a result of sampling errors or because of poorly defined measurement methods. Consider the failure of the Gallup Poll forecast in the Landon–Roosevelt presidential contest in 1936. It used a telephone survey and therefore, in a day when telephones were less common than they are now, biased its sample towards a higher income group favoring the Republican candidate. Consider also failure of the automobile purchasing surveys in the early 1950's which showed that people wanted smaller, less brightly finished cars when in fact they continued to buy longer, more exotic vehicles for many more years. The defect in the first instance is a sampling error, and, in the second, inadequate exactness in the defining of the quantity to be measured. Methods are available for the programming of such observational type experiments by statistical methods to minimize the chance of errors, and many texts are devoted exclusively to this topic. (See, for example, Reference 3 in the Bibliography.)

(c) *Statistical Fluctuation*

It is probably true to say that the influence of statistical fluctuation does not vary a great deal. It is always too much. This seems to be true whether an experimenter is studying the influence of fluorine on tooth decay where the spread of results may be comparable with the size of the effect, or a metrologist is losing sleep over the ninth significant figure in a wavelength measurement.

The influence of statistical fluctuation has been considered in the last chapter, and it must be apparent that continued duplication of readings is unrewarding in a measurement of low intrinsic precision. One way of attacking this problem is to measure differences rather than absolute values. The dangers of obtaining a measurement as the difference between two absolute values has been pointed out earlier and, if it is possible to measure the difference itself, much higher precision can be achieved.

The advantages of this are obvious in mere numerical considerations, before any experimental problems are discussed. If one has a resistor of around 1000 ohm resistance (suppose 1124 ohms) and wishes to know its resistance to within 1 ohm it is necessary to measure it with a precision better than $\frac{1}{10}$ per cent. If however we have a standard resistor of value 1000 ohms, and arrange that we measure the *difference* between the two resistances, we must measure this quantity (124 ohms) to a precision of about 1 per cent only to achieve our objective. The use of "beats" for frequency measurement is another example of this technique. This use of a standard for comparison purposes is especially valuable when some perturbing influence is at work on both the sample and the reference simultaneously. Thus the effect of temperature changes on the leads to a platinum resistance thermometer is eliminated by subtracting the resistance changes of a dummy pair of leads subject to the same temperature changes, and in instruments measuring the optical transmittance of liquids, changes in source intensity and detector sensitivity are eliminated by alternating the light path between the sample and a vacant cell, and measuring the *difference* between the two intensities. These instrumental methods are really a process of continued calibration, but the same approach is found in other sciences where it is desired to subtract out the effect of perturbing influences. The botanist will study the effect of fertilizer on wheat not by taking a wheat patch, adding fertilizer and measuring the resulting crop. The perturbing effects of rainfall, temperature, sunshine, etc. are much too great for this approach to be effective, and so the fertilized product is compared with another plot of wheat called a "control," grown under exactly the same conditions except for the fertilizer. The difference between the two plots of wheat is the effect of the fertilizer, which can therefore be measured in isolation from unwanted influences. Another example of common occurrence is particle counting in nuclear physics. Here the perturbation is the addition of unwanted counts from cosmic rays and a control experiment is necessary in which the "background" counts are measured by themselves for future subtraction from the main measured value. Note that all these experimental procedures have one thing in common; the difference which is measured is caused by the factor under investigation (the difference between the two plots of wheat is the fertilizer, the difference between the two counting experiments is the presence of the radioactive source, etc.) and everything else is held constant. Make sure that the experiment is designed to do that and not something else: it is rather easier than one might think to try the physics laboratory equivalent of investigating the effect of fertilizer by varying the rainfall. (See also on page 114 another example of this technique in the "sample in–sample out" method.) Such "control" type experiments are especially valuable when the control and the specimen are subject to the same fluctuations. Even when they are not, however, a control or standard is still useful, as mentioned above, for transforming a $\frac{1}{10}$ per cent measurement into a 1 per cent measurement.

Many measurements in science are of the type in which the quantity desired is a function of some variable e.g., electrical resistance as a function of temperature. In many cases the important quantity is this variation itself and the absolute value is less important. This situation is, once more, a difference type problem. Suppose a resistor varies between 1000 and 1100 ohms over a certain range and the variation is required to 1 per cent. Clearly one should measure the *change* in resistance directly, either with respect to the initial resistance itself or in comparison with some fixed standard of suitable value.

Difference methods are very powerful and should be considered whenever possible in designing experiments. The only requirement in realizing the potentialities of the method is the reliability and stability of the standard quantity. There is no point in attempting to measure a frequency to 1 cps using beats, if the reference oscillater is unstable to 5 cps.

A special case arises when, in such comparison-type measurements, the reference value can be controlled so that the measuring instrument reads zero at the time the measurement is being made. This is the case, for example in the Wheatstone bridge where the reference resistor is adjusted to give zero current in the galvanometer, in the potentiometer when the unknown emf is balanced against a value obtainable from the slide wire, and in the optical example mentioned on p. 85 in which the reference intensity can be controlled so that the difference being measured is zero. In all these cases one very important advantage is secured and that is that the final reading is independent of the characteristic of the detecting instrument and possibly also of other components in the system. This constitutes the so-called "nul" measurement. It is a very valuable method because it eliminates problems of calibration of instruments like meters and amplifiers, and replaces them by the preferable problem of calibration of reference resistors, potentiometer slide wires or optical shutters. It may also eliminate problems of source stability as, for example, in the Wheatstone bridge for which, at the balance point, non-constancy of the supply battery is unimportant.

5 Experiment Planning

The conclusion from the last chapter is that an experiment is a much more sophisticated procedure than an isolated measurement, because it is the investigation of the properties of a physical system. The results of such an investigation must necessarily be expressed in terms of the commonly accepted ideas concerning the nature of the physical system. The conduct of the experiment will therefore depend very largely on how much of this background material is available, but in all cases the essence of good experimenting is planning. It is courting disaster to have the (not uncommon) attitude of rushing through the measurements and worrying later about what to do with them. The experimenter should remember this and take time, before starting an experiment in the laboratory, to analyze his experiment in the ways to be described, and to lay out his analysis, and the consequent measurement and calculation program, formally and neatly in his laboratory note book. We have been stressing the wide range of experimental activity, but from now on we shall be restricted more to the practices of the physics laboratory.

5.1 Precision of Measurement

Whatever the nature of the experiment, it will be constructed out of measurements, and we must consider these first. The conduct of the experiment will be determined very largely by the precision which is required. The instruments used to measure individual quantities, and the whole method of measurement will depend on it. Thus every experiment should be considered in the light of some desired precision for the answer. This quantity should be chosen realistically, since too optimistic a value will very quickly lead to too great complexity. A desire to measure e/m for electrons to $\frac{1}{10}$ per cent in an elementary laboratory would almost certainly lead to disappointment. Once the experimenter knows, in general terms the precision for which he is aiming, he must then turn his attention in turn to each of the quantities involved in the measurement. Remember at this stage the important distinction between readings whose precision is limited by statistical fluctuation and those whose precision is limited by the measuring scale. The only way to distinguish between those two types is to try the measurement to see. Does repetition give the same scale reading or not? Once this point has been settled is the apparent value of the uncertainty acceptable or not? If not, then some improvement is required. If the precision is scale-limited, the acquisition of a more precise instrument is indicated. If such is not available, then a lower limit to the uncertainty of measurement is already set. If the uncertainty of the reading proves to be statistical in nature, obtain an estimate of the standard deviation using, say, 10 readings. Does the precision thus calculated appear adequate or not?

If not, can the precision be improved either by taking more readings or must the measurement procedure itself be improved? Remember from page 33 how the standard deviation of the mean involves $\sqrt{n}$. Consequently, if 10 readings suggest that the precision must be improved by a factor of 10 (the fluctuations may be 10 per cent, when the experimenter wants 1 per cent), the number of readings must be increased by a factor of 100. This is an undesirable method of improving precision and, in this case, some improved measurement procedure is called for. [Refer to Sec. 4.2 (c).] If this is not possible then, once again, a lower limit to the uncertainty has been set by the time and resources available for duplication of the readings. The whole skill in experiment design lies in the optimum choice of the above procedures to give maximum precision, taking into account the available resources of equipment, time and money.

At this stage use the methods of Chapter 3 to find out how each elementary uncertainty is propagated through to the final answer. This will give an estimate of the over-all uncertainty to be found in the experiment, and enable the experimenter to identify those quantities which dominate the construction of the final uncertainty either because of poor intrinsic precision or because they are raised to a high power. The experimenter is now in the position of having considered fully the measurement aspect of his experiment. He has either secured the apparatus necessary to achieve the required precision, or else he has obtained an estimate of the precision to which he is limited by the resources available. The subsequent planning of the experiment will depend on how much theoretical material he has to suggest the behavior of his system.

5.2 Experimenting with No Background

The circumstances here are that we are presented with a physical situation about whose characteristics we know virtually nothing. The situation might involve the strength of a concrete mix which is dependent on the proportions of its constituents, the angular dependence of protons scattered from an atomic nucleus, or the efficiency of a gasoline engine which is known to depend on speed and fuel mixture richness. The task is to investigate the system. This investigation can have two aims. First, the experimental results will describe the system to anyone else. Once the curve of concrete strength vs. proportion of sand has been established, other people can then make up concrete to have a given strength by using the prescribed quantity of sand. Secondly, the results will serve as a guide to a theoretician who wishes to construct a model of the situation. In engineering the first aim is more common, in physics the second.

In both cases, the design of the experiment starts with a selection of the variables. There is usually one obvious variable which we can select to be the main measured quantity (the strength of the concrete, the intensity of the scattered proton beam or the efficiency of the engine). This is called the *dependent variable*. This quantity is probably influenced by a whole host of other factors and the next task is to select those whose influence we wish to measure (proportions of sand, angle of proton scattering, engine speed, and mixture). These are called *independent variables* since we are able to choose their values at will, thus determining through the properties of the system the value of the dependent variable. Thus, it is important to ensure that, once the independent variables are chosen, all other factors which may influence the dependent variable be held closely constant (e.g., setting time of the concrete, energy of the incident proton beam, engine intake air temperature). If everything starts varying simultaneously we shall find it very difficult to understand the results.

Once the variables are selected and the measurement techniques established according to the precision desired, as described in Sec. 5.1, the measurement program can be constructed. The form of the measurement program will obviously be the measurement of the dependent variable for several values of the independent variable, the aim being to learn as much as possible about the behavior of the system. The values of this independent variable should cover as wide a range as possible, and obviously the number of values should be as large as possible. The actual number will be predetermined by the time available for the whole experiment and the time required for each measurement. If there is more than one independent variable, the technique is to choose one of the independent variables and to hold it at a constant value while one measures the variation of the dependent variable on the other. Thereupon another constant value is chosen for the variable and the process repeated. Thus, we would have for the gasoline engine a series of curves of efficiency vs. engine speed, each curve referring to a certain value of the fuel mixture richness as shown in Fig. 5.1. If anyone wanted to have the variation of efficiency with mixture richness for a given speed, he would have to read the values off the graph along such a line as *ABCD* and replot the values accordingly. A variable which is held constant while an experiment is in progress

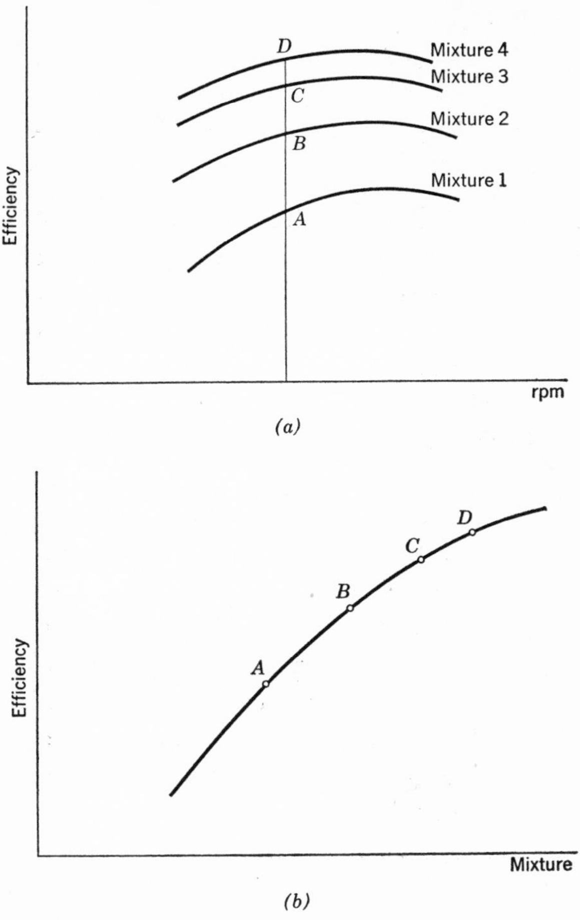

Fig. 5.1 Engine efficiency as a function of mixture and rpm.

and given a few discrete values, as described above, could be termed a "subsidiary variable" as opposed to the main or principal variables.

The end point of such an investigation is a curve or a set of curves. If this is an empirical study it may suffice merely to present the curves saying, in effect, "Regardless of what is really going on, this is what the system actually does." If the work is merely a prelude to theoretical thought, further analysis is needed and this will be described in Chapter 6.

Although it is usually taken as an axiom of good experimenting in the scientific laboratory to hold all independent variables constant except the one under investigation, this commonly turns out to be a counsel of perfection in more complex systems, e.g., in industry. Here the variables may be interdependent so that the effect of varying one may be to alter all the others which the experimenter had hoped to keep constant. As an elementary example, consider an experiment to evaluate the static constants of a pentode tube. The plate and the screen grid are connected through suitable potentiometers to a power supply, and a separate grid supply is provided. The aim of the experiment might be to measure a defined parameter of the tube such as the variation of plate current with grid potential for constant screen potential. One would do this by setting the grid and screen potentials to the required initial values and reading the plate current. The next step would be to change the grid potential to the next required value and read the plate current again. However, the experimenter is very likely to find that the altered plate current has changed the source voltage so that the screen potential is altered from its original value. In other words the supposedly independent variables are not independent. This is a trivial example, of course, because it is a simple matter to reset the required screen potential. However, the experiment may be on a rather larger scale such as the rate of production in industrial continuous flow processes in which the flow rate depends on the reaction rate which depends on the temperature which itself depends on the flow rate which .... , or a problem in meteorology in which the variables, wind velocity, pressure, temperature, humidity, height, etc. may be all interdependent to a high degree.

All one can say about systems such as these is that the dependent variable is a function of the independent variables and various constants

$$Z = f(x_1, x_2, x_3, \ldots, a_1, a_2, a_3, \ldots)$$

If one wants information about such a system, one wants the values of the constants a_1, a_2, etc. and, to determine them, all one can do is make observations of the required quantity under various sets of circumstances with the values of the various variables falling where they may. With many variables such a procedure is sure to lead to chaos unless some system of selecting the observations is used. The methods which have been developed for choosing a measurement scheme, which will yield the maximum amount of information for a given experimental effort, form a very important part of experiment design, they are of widespread use in science and technology, and much of the literature is devoted to them. They are not considered here further because the physics laboratory normally enjoys the luxury of controllable variables. For further information the reader is referred to References 3, 11, 18, and 19 in the Bibliography.

5.3 Dimensional Analysis

Even if no complete theory of a physical phenomenon exists, it is still possible to obtain very useful guidance to the performance of an experiment by the method of dimensional analysis. The "dimensions" of a physical (mechanical) quantity are its expression in terms of the elementary quantities of mass, length, and time, denoted by M, L, and T. Thus, velocity has dimensions LT^{-1}, acceleration LT^{-2}, density ML^{-3}, force (equal to mass $\times$ acceleration) MLT^{-2}, work (equal to force $\times$ distance) ML^2T^{-2}, etc.

The principle used in dimensional analysis is that the dimensions on each side of an equation must match. Thus, if g is known to be dependent on the length and period of a pendulum, it is obvious that the only way in which the LT^{-2} of the acceleration can be balanced on the other side is by having the length to the first power (to give the L) and the period squared (to provide T^{-2}). We can thus say immediately that, whatever the final theoretical form for the equation it must have the structure

$$g = (\text{dimensionless constant}) \times \frac{\text{length}}{\text{period}^2}$$

Note that the treatment can give no information about dimensionless quantities (pure numbers, π, etc.) and so we must always add in such a possibility to the form of an equation obtained by dimensional analysis.

The general method is as follows: Consider a quantity z which is assumed to be a function of variables x, y, etc. Write the relation in the form

$$z \propto x^a y^b \ldots$$

where a and b represent the numerical powers to which x and y have to be raised. The values of a and b will then be found by writing down the dimensions of the right-hand side in terms of the dimensions of x and y and the powers

a and b, and writing down the condition that the total power of M on the right-hand side must be the same as that known for z, and similarly for L and T. Three simultaneous equations result which enable values for a, b, etc. to be calculated.

For example, consider the velocity v of transverse waves in a string. We might guess that this velocity is governed by the tension T in the string and the mass per unit length m.

Let us write

$$v \propto T^a m^b$$

dimensions of v, $\quad LT^{-1}$

dimensions of T, $\quad$ (force), MLT^{-2}

dimensions of m, $\quad$ (mass/unit length), ML^{-1}

Therefore,
$$\begin{aligned} LT^{-1} &= (MLT^{-2})^a(ML^{-1})^b \\ &= M^{a+b}L^{a-b}T^{-2a} \end{aligned}$$

Therefore, by comparing powers of

$$\begin{aligned} M, \quad 0 &= a + b \\ L, \quad 1 &= a - b \\ T, \quad -1 &= -2a \end{aligned}$$

of which the solutions are obviously

$$a = \tfrac{1}{2}, \qquad b = -\tfrac{1}{2}$$

so that we can write

$$v = (\text{dimensionless constant}) \times \sqrt{T/m}$$

Such a treatment is very valuable, for it provides, even in the absence of a detailed fundamental theory, a prediction regarding the properties of the system. This is available for experimental investigation and, if our original guess regarding the factors contributing to v was correct, the prediction will be verified. If experiment shows a discrepancy, then we must look again at our original guess. Notice that we obtained three equations for only two unknowns. The condition, therefore, was really overdetermined and we were very fortunate that the equations were consistent. Had they not been consistent we would have known immediately that our guess regarding the constituents of v was wrong.

Powerful as this method is, difficulties will obviously arise when the quantity under discussion is a function of more than three variables. Thus, we shall have more than three unknown powers but only three equations from which to determine them. In this case a unique solution is not possible but a partial solution may be found in terms of combinations of some of the variables.

For example, consider the flow rate Q of fluid of viscosity coefficient η through a tube of radius r and length l under a pressure difference p. We may suggest a relation

$$Q \propto p^a l^b \eta^c r^d$$

The dimensions of the quantities are as follows:

Q, volume per unit time, $\quad L^3T^{-1}$

p, force per unit area, $\quad MLT^{-2} \cdot L^{-2} = ML^{-1}T^{-2}$

l, $\quad L$

η, viscosity coefficient is defined as a force per unit area per unit velocity gradient,

$$(MLT^{-2})(L^2)^{-1}(LT^{-1} \cdot L^{-1})^{-1} = ML^{-1}T^{-1}$$

r, $\quad L$

Therefore, $\quad L^3T^{-1} = (ML^{-1}T^{-2})^a L^b (ML^{-1}T^{-1})^c L^d$

Comparing powers of

$$M, \qquad 0 = a + c$$
$$L, \qquad 3 = -a + b - c + d$$
$$T, \qquad -1 = -2a - c$$

Here we have four unknowns and only three equations so that, in general, a complete solution is not possible. We can find part of it, however, for obviously

$$a = 1$$

and

$$c = -1$$

We must have, therefore,

$$Q \propto \frac{p}{\eta}$$

The remaining part of the solution can be written only as

$$b + d = 3$$

If we write this

$$d = 3 - b$$

we can see that Q must contain the product r^3/r^b. It also contained l^b so that we can write finally

$$Q \propto \frac{p}{\eta} r^3 \left(\frac{l}{r}\right)^b$$

and this is as far as dimensional analysis can take us towards the complete solution. However, even this partial solution would be of enormous assistance as a guide to experimenting in a situation in which no fundamental theory existed.

Dimensional analysis can be extended to cover thermal and electrical quantities, but ambiguities arise which require special consideration. Treatments will be found in the standard texts on heat and electricity.

5.4 Experimenting with a Theoretical Background

The situation is rather different when there is a theoretical background to the experiment. This theoretical background may range all the way from a mere suggestion about how the system might behave to a well established and highly developed theory. In many cases where the theory is more highly developed the quantity which is required to be measured may be more complicated than the simple observational properties found in empirical work (the strength of the concrete or the efficiency of the engine), and may be defined only in terms of the theory. Such a quantity would be the constant of gravitation G defined by the equation for the gravitational force F between two masses m_1 and m_2 at distance r

$$F = G\frac{m_1 m_2}{r^2}$$

This situation uses the inverse square law of gravitational force as a model and the constant G is meaningless unless interpreted according to this model. The point is that there is no a priori guarantee that the conditions described in the theory match those found in the apparatus. A discrepancy which exists between experiment and theory can range from the simplest of systematic errors such as an unnoticed instrumental zero error to definitive evidence on the status of great theories.

The theory is a deduction from an idealized model (a pendulum is a point mass suspended from a weightless, inextensible string, etc.) and the experiment is conducted on actual bits of apparatus in the laboratory. The extent to which it is valid to interpret the results of the experiment in terms of the theory is the responsibility of the experimenter to determine. In general, such proof of validity is absolutely necessary before any experimental result can be accepted. This is why, even in situations where the topic is familiar (resistivity and a Wheatstone bridge), the experimenter should keep his theoretical background in mind. (Some day our Wheatstone bridge operator might be given a piece of non-ohmic material to measure.) Note that we do not say the theory is "wrong" or the experiment is "wrong." It is merely a matter of whether the idealized conditions of the model approach sufficiently closely the conditions of the experiment. If they do not, the behavior of the actual system will deviate from that of the ideal system and a systematic error may be introduced. The question of the importance of this systematic error depends entirely on the precision of the experiment, and only the experiment itself will enable one to judge whether the correspondence between theory and experiment is adequately close. It should be unnecessary to mention that any superficially obvious mismatch should be corrected before further work is done. (If the pendulum string is noticed to be really stretchy, the experimenter either obtains another piece of string or else develops the theory of pendulums with elastic strings.)

In general, then, the question of whether the theory and experiment are adequately in correspondence is to be settled by observation of the behavior of the system itself, i.e., is the experiment working out as expected or not? The clue as to how this is to be determined lies in this expectation because any theoretical result can be regarded as a prediction. Thus the expression for the flow rate Q of fluid of viscosity η along a pipe of radius a and length l under pressure p

$$Q = \frac{\pi p a^4}{8\eta l}$$

is a prediction that

$$Q \propto p$$
$$Q \propto l^{-1}$$
$$Q \propto a^4$$

Clearly it is impossible to judge from one spot reading whether these predictions are fulfilled or not; the only way is to take a series of readings with different values of the variables. Once again we have confirmation of the uselessness of isolated spot readings. Only if these ranges of variables have been covered as widely as possible, and the behavior found to be as expected, can the experimenter claim that his theory and experiment are compatible within the precision of measurement and that, consequently, his final answer is a valid one. It is true that not all experimental defects can be detected in this way since a discrepancy may be such as to affect all readings equally, and this does make things difficult. However, the covering of a range of values of variables goes very far towards elimination of this source of uncertainty. The only way in which one can improve the chances of eliminating *all* systematic errors is by comparison of measurements of the required quantity using completely different measurement methods.

Once the range of readings has been taken, the question arises of the comparison between the observed variation and that predicted. A column of figures on paper is obviously almost useless since no one can say by looking at them that the variation is, say, a fourth power. The only available way of judging is by graphical methods, because if the results are laid out in pictorial form, it can become obvious whether the predictions are fulfilled or not. However, it would not suffice merely to plot Q vs. a if one had done an experiment with a lot of pipes of differing a while holding p and l constant. The result of such a procedure would be a curve, and no one can judge visually whether such a curve is a fourth power or not. The only curve which has the convenient property of permitting visual judgment is a straight line, and so the task is to plot the observations in such a way that they would form a straight line if they obey the theoretical predictions. Clearly in this case one would calculate the quantity a^4 for all the values of a and plot Q vs. a^4. If the theory and apparatus are in correspondence, the points will form a straight line and any lack of correspondence will be revealed by a departure of the points from linearity along all or part of the series. The experimenter then knows the range of validity of his experiment. He will therefore calculate his required answer from the points falling on the straight line, rejecting those which deviate. This is not dishonest selection of observations but merely a restriction of the calculations to the range of validity within which his quantity is defined. For example, a plot of fluid flow Q vs. p in the previous example might give a result such as shown in Fig. 5.2. The experimenter could then restrict his calculation of his required quantity (probably η) to the linear region, and reject the discrepant points. He could be sure that these last correspond to conditions outside the framework of his theory, whether or not he knew anything about the onset of turbulence. Another common form of discrepancy is a shifted origin. A set of points which should lie on a line passing through the origin

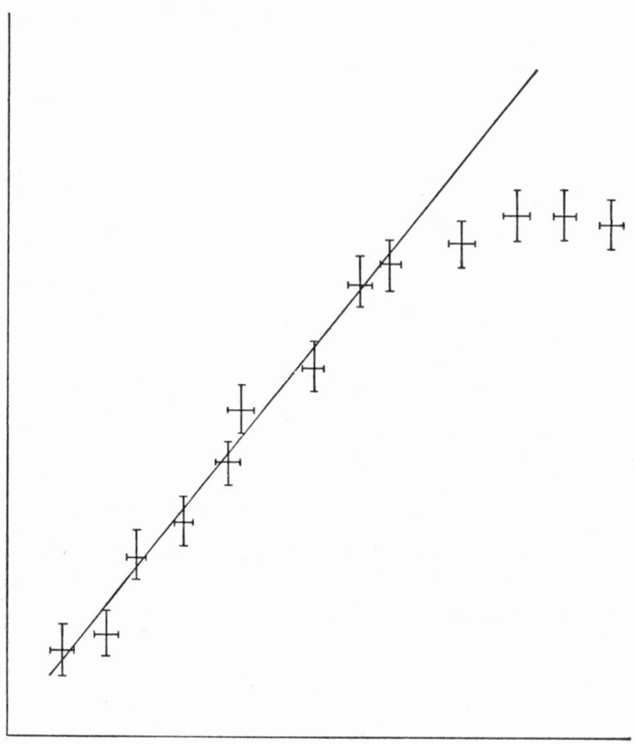

Fig. 5.2 The identification of the range of validity of an experiment.

might actually give an intercept on one axis but here, again, a mode of computation using the slope can be selected so that no systematic error in the answer is permitted. It may not at this stage be possible to say in any particular case what the discrepancy between theory and apparatus is, but it does not matter. It is sufficient for the moment to have it revealed, so that it is not permitted to introduce errors into the answer.

So far we have described the graph merely as an indicator of the validity of the experiment. There are many more advantages to graphical analysis. These advantages and the methods of working with graphs must now be described.

5.5 Graphical Analysis

It should be noted here that graphs are of two general types—pictorial and computational. The first type is used to illustrate the behavior of a physical system (period vs. point of suspension for a physical pendulum, thermionic current vs. plate potential for a thermionic diode), being merely a pictorial description of the properties of the system, and these are commonly found in texts. They are very pretty but are no help in judging an experiment or (in general) in computing an answer. The second type, which is almost always a straight line, has the purpose of assisting the evaluation of the experiment and the computation of an answer, and this is the type with which we are almost exclusively concerned. In all cases the advantages of graphical presentations of experimental results are overwhelming. They include: verification of validity of the experiment, as has been described above; ease of calculation of final answers, as will be described below; a check on the over-all precision of the experiment since the uncertainties in each observation will be revealed by scatter of the points.

In order to extract all the information from the observations it is necessary to plot not only the measured values but their range of uncertainty (ways of doing this will be suggested later on page 128). Only if the range of uncertainty is plotted on the graph will the significance of any deviation be apparent and will the over-all uncertainty of the experiment be evident.

In considering the linear graph as an aid to computation we must consider what information is available from a graph once it is drawn. Two pieces of information are available, which can be taken as a slope and intercept or as two intercepts. Clearly the object of the analysis of the equation into linear form is to cast the unknown (or unknowns if there are two) into the role of constants, i.e., slope or intercept, while retaining only measured quantities in the variables. There is no standard method for doing this and a unique solution does not necessarily exist. The best way of attacking the problem is to keep clearly in mind the straight line equation

$$y = mx + b$$

while considering the relation relevant to the experiment. The best way of illustrating the process (known as the "rectification of the curve") is by examples.

(a) *Ohm's law*

$$V = IR$$

R = resistance of resistor (constant and unknown)
V = potential (dependent variable)
I = current (independent variable)

This is already linear in the form

$$y = \text{slope} \times x$$

where y is V, x is I and the slope is R. Consequently a set of measured V and I will enable a straight line to be drawn and the slope measured, thus yielding R.

(b) *Gas law*

$$pv = RT$$

$$\left.\begin{aligned} p &= \text{pressure} \\ v &= \text{volume} \end{aligned}\right\} \text{ of one mole of gas}$$

T = temperature
R = gas constant per mole

If T has been held constant, the equation reads

$$pv = \text{constant}$$

This can be rewritten

$$p = \text{constant}\ \frac{1}{v}$$

and, once again, we have straight line form with p and $1/v$ as y and x respectively. A measurement of the slope will yield a value for the gas constant R, provided the temperature is known.

(c) *Fluid Flow*

$$Q = \frac{p\pi a^4}{8\eta l}$$

If the radius a and length l are kept constant, the equation has the form

$$Q = \frac{\pi a^4}{8\eta l} p$$

and is already in straight line form with y as Q, x as p and the slope as $\pi a^4/8\eta l$.

If a and l are known then η can be calculated. If a and l are also variable, a number of methods could be used. Q vs. p could be plotted as a series of lines, each one referring to discrete values of a and l (see page 92). This describes the system but does not check the a^4 and l dependence. One could plot Q vs. a^4 for constant p and l, or Q vs. l for constant p and a^4 but a rather neater way of summarizing the results would be to use a compound variable. The equation can be written

$$Q = \frac{\pi}{8\eta}\frac{pa^4}{l}$$

where y is Q, x is pa^4/l and the slope is $\pi/8\eta$ (for a precaution regarding this case see page 128). The use of such a compound variable is perfectly valid and, while optional in this case, may facilitate the assessment of complicated sets of observations.

(d) *The Compound Pendulum*

$$T = 2\pi\sqrt{(h^2 + k^2)/gh}$$

T = period (dependent variable)
h = distance from CG to point of support (independent variable)
g = gravitational acceleration (constant and unknown)
k = radius of gyration about CG (constant and unknown)

The treatment of this is not obvious, but it is clearly impossible to place it in the required linear form where y and x are functions of h and T singly. An analysis into compound variables is, however, possible

$$T^2 = 4\pi^2\frac{h^2 + k^2}{gh}$$

$$T^2h = 4\pi^2(h^2 + k^2)$$

$$h^2 = \frac{g}{4\pi^2}T^2h - k^2$$

Compare with $$y = mx + b$$

to give a linear form with y as h^2, x as T^2h, the slope as $g/4\pi^2$ and the intercept as k^2.

This example has been chosen specifically since it illustrates very clearly the superiority of linear analysis over other methods. The graph of T vs. h which can be obtained is shown in Fig. 5.3. It turns out that k can be obtained from

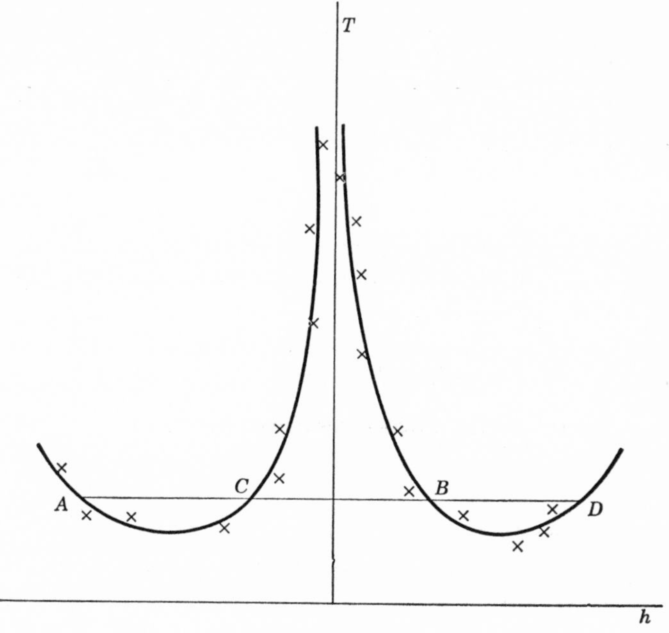

Fig. 5.3 The behavior of the compound pendulum.

the lengths of the intercepts AB and CD. If g is required, it has to be obtained as a calculation from this value of k. The advantages of the linear analysis are clear. The T vs. h graph gives no check on the performance of the experiment except in the most general terms. No reliable estimate of the uncertainty of the final answer can be obtained from this graph, but the over-all uncertainty can be obtained readily from the linear graph (see page 131). The use of an intercept at such a low angle, as illustrated in Fig. 5.3, is very unreliable, since small changes in orientation of the lines can make large changes in the length of the intercepted portion. On the other hand, the slope of a linear graph can be determined very reliably. The answer using the intercept method is determined solely by a few points in the vicinity of the intercepts, and the value of all the other points is not realized. When drawing a straight line, however, all the points contribute towards the choice of the line. Lastly, the linear graph gives g and k from independent measurements on the graph while, in the other method, any inaccuracy in the value of k is propagated automatically into the value of g.

(e) *Logarithmic Functions*

Many physical processes are of the form

$$Q = Q_0 e^{-a/T}$$

where Q is a measured variable, T is the temperature and Q_0 and a are constants. This can be linearized by taking logs to the base e

$$\log_e Q = \log_e Q_0 - \frac{a}{T}$$

Thus, log Q plotted vs. $1/T$ will yield a straight line of slope a and intercept $\log_e Q_0$. Note that if logs are taken to the base 10, only the intercept is affected, and this is a convenience if only the slope is to be measured.

Other more complicated functions are commonly found but, in all cases, some kind of linear analysis can be achieved, provided one is prepared to accept compound variables. Such a linear analysis should always be attempted. If there is any suggestion at all that a particular function can be expected in an experiment, try it. Any clue is better than none and even if the suggested function is wildly wrong, the nature of the discrepancy will almost certainly be useful as a guide to further theoretical thought. At the other end of the scale *all* expected behavior should be checked, even when the experimenter feels sure that everything is working as it should. If the current should be proportional to the voltage, make sure it is. When the experimenter is describing his experiment after it is all over, someone is sure to ask, "Did you check?"

5.6 Experiment Analysis and Design

The task of approaching an experiment is therefore a little more complicated than throwing together the pieces of apparatus provided, reading some scale just because it had been left beside the apparatus, and deciding to postpone worrying about what to do with the measurements until after it is all over. The experimenter should remember that an experiment is something one does because it is the only way of obtaining the information one needs. Therefore, the act of observation should be the consequence of planned necessity. The point about an experiment is not so much the observations themselves as why they were taken and what the observer does with them.

There is no justification, therefore, in starting to take observations unless the experiment has been completely analyzed, the mode of computation selected, and a program of measurement drawn up. The steps in this process have already been described but we wish to summarize the procedure.

(a) *Selection of Variables*

This will usually be obvious. The experimenter will have been given the task of measuring a specific quantity as a function of other specified variables (e.g., triode plate current as a function of grid potential and plate potential), or else the required answer will be known from theoretical grounds to involve certain other measurements. Decide which quantities are to be the principal variables and which are to be the subsidiary variables, and try to make sure that you do not inadvertently change two variables at once.

(b) *Graphical Analysis*

If there is any clue at all to graphical analysis of the problem in straight line form, use it. The process of putting the equation in straight line form will determine the mode of computation since the unknown will appear in a slope or intercept. Remember that the aim of the experiment is twofold, first, a check on the validity of the experiment and second, a calculation of the answer. Make sure that both requirements are met.

(c) *Experimental Precision*

Make trial measurements of the various quantities to consider the precision as outlined in Sec. 5.1. When a suitable compromise has been worked out between the precision desired and that attainable using the available apparatus, make an estimate of the over-all precision of the experiment which is likely to be achieved. This will serve as a very useful guide in the conduct of the experiment.

(d) *Measurement Program*

Make up a complete schedule of measurement for all the quantities shown by the experiment analysis to require measurement. The mode of measurement (e.g., micrometer or meter stick) will already have been selected and so the measurement program must allow for adequate duplication of those readings which require statistical treatment. In general, take care to ensure that adequate attention is given to the quantity in the experiment which has the dominant influence on the uncertainty, but do not waste time over others. If one quantity in an experiment cannot be measured with an uncertainty less than 5 per cent, it is a waste of effort to spend time in an attempt to reduce the uncertainty of another, equally significant, quantity from $\frac{1}{2}$ per cent to $\frac{1}{4}$ per cent.

The range of values to be taken will be determined by instrumental limitations, and the number of different values by the time available. This is the time to consider instrument and equipment ratings. The rather expensive smell of charring insulation coming from a 1 ohm standard resistor can easily be avoided by noting beforehand that a maximum current value of 1 amp is clearly stamped on the casing. This matter of ratings is obvious in instruments with scales (ammeters, pressure gauges, etc.) but is not less important in other components like resistors where no scale is involved. In such a case, always look carefully for the rating value marked on the component, and adjust the measurement program accordingly.

The measurement program should allow for all factors which may act as a check on the progress of the experiment. For example, all reversible quantities should be read both ways, e.g., measure elastic deflections both when loading and unloading, use both direct and reversed currents if possible, take measurements both when heating and cooling, read both ends of circular scales, etc. Each of these procedures will detect some kind of systematic defect. In elastic deflections measurements, unloading in addition to loading will check that an elastic limit has not been exceeded. In an experiment using the magnetic field of coils, the use of both direct and reversed currents will check the uniformity of the magnetic fields. In temperature variation experiments, the problem is always to ensure that temperature equilibrium has been achieved. Only a cooling as well as a heating experiment will be convincing. In general, one can be sure that one is dealing with equilibrium conditions only if the experiment works as well backwards as forwards.

Another very useful check is the "sample in–sample out" procedure as described on page 85. This means that the characteristics of the system are studied both with and without the element whose properties are desired. The difference between the two observations must be due to the object under test. This procedure thus makes the experiment act as its own control, and is especially useful when there are a lot of perturbing influences. For example, if we are measuring the thermionic current in a diode it is wise to check the current when the filament is not heated. Any current which is observed must be due to leakage, and thus constitutes a systematic error in all the actual observations. Anyone who doubts the value of such nul-measurement checks should follow the advice of Wilson (Reference 18 in the Bibliography) and contemplate the statement, "It has been conclusively demonstrated by hundreds of experiments that the beating of tom-toms will restore the sun after an eclipse."

In constructing a measurement program it is usually worth

considering the topic known as randomization of the readings. Consider that a series of different alloys are to be given a plastic deformation test using a drop hammer. The idea is to study the variation of deformation with alloying concentration, but it may happen that, as the drop hammer works during the course of the tests, its lubrication becomes less sticky and the actual impulse delivered to the specimen is increased. If this effect were not suspected, a serious contribution to the observed variation is provided. Now, if the alloy specimens are used in the expected sequence of increasing hardness, this discrepancy could probably go unnoticed since it is a smooth variation added to another smooth variation. If, however, the samples are tested in random order, the discrepancy will be detectable. The error has not been removed but it has been changed from a systematic error, difficult to detect, to an easily visible scatter of the points. This discovery of an unsuspected source of uncertainty can then be used to eliminate it. Note that this method is equivalent to the method, already mentioned, of taking readings "going both ways": they both detect progressive systematic defects. If complete control in the experiment is possible the first method is preferable, but if this is not possible (as in the case of destructive testing where only one sample of each type is available) randomization is recommended.

It will enormously simplify computation later on if the measurement program is laid out in the form of a table, which incorporates all the stages of the future calculation, both of the best values and their uncertainties. Consider, for example, the problem of a mass m oscillating at the end of a spring. The period T of vibration can be written

$$T = 2\pi\sqrt{m/k}$$

where k is a constant of the spring which it is desired to measure.

Straight line analysis gives

$$T^2 = \frac{4\pi^2}{k} m$$

so that we plot T^2 vertically and m horizontally to obtain k from the slope. The measurement program would then be as follows:

m	AUm	T	AUT	RUT	T^2	RUT^2	AUT^2

where AU and RU mean absolute uncertainty and relative uncertainty respectively. The reasons for the format above are as follows: m is obvious and we shall need AUm for plotting on the graph; T is the measured dependent variable but we need T^2 for the graph; we need AUT^2 for plotting and it must be calculated from the AUT (which we assume to be known) through the medium of the relative uncertainties of T and T^2. In this simple case it would have been possible to calculate AUT^2 as $2T(AUT)$ instead of using the relative uncertainties. This would have saved a column of calculation but it is commonly very useful to have the relative uncertainties available, especially when dealing with compound variables (e.g., quotients). Each case should be treated on its own merits.

In actual professional work where money, labor and equipment may all be involved to a large degree, the onus is on the experimenter to perform his experiment so as to secure the maximum yield of results with the resources available. The highest efficiency can only be achieved by adequate preparatory thought. If the analysis and measurement program construction above is completed, the taking of the observations will be a profitable consequence of reasoned necessity rather than optimistic guesswork, and the possibility of omitting necessary and probably irretrievable measurements is eliminated. The personal strain of observation is much reduced and this is conducive to good observing, so that the over-all chances of success are enormously improved. Without such planning an observational process cannot be granted the title of an experiment and reduces itself to an undignified scrabbling for data in the hope that something will prove useful.

PROBLEMS

1. A scientist claims that the terminal velocity of fall of a parachutist is dependent only on the mass of the parachutist and the acceleration due to gravity. Is it worth while setting up an experiment to check this?

2. The range of a projectile fired with velocity v at angle α to the horizontal may depend on its mass, the velocity, the angle and the gravitational acceleration. Find the form of the function.

3. The pressure inside a soap bubble is known to depend on the surface tension of the material and the radius of the bubble. What is the nature of the dependence?

4. The period of a torsion pendulum is a function of the rigidity constant (torque/unit angular deflection) of the support and of the moment of inertia of the oscillating body. What is the form of the function?

5. The deflection of a beam of circular cross section supported at the ends and loaded in the middle is dependent on the loading force, the length between the supports, the radius of the beam and Young's modulus of the material. Deduce the nature of the dependence.

In all the following problems state the variables or combination of variables which should be plotted to check the suggested variation and state how the unknown may be found (slope, intercept, etc.).

6. The position of a body starting from rest and subject to a uniform acceleration is described by

$$s = \tfrac{1}{2} at^2$$

s and t are measured variables. Determine a.

7. The fundamental frequency of vibration of a string is given by

$$n = \frac{1}{2l} \sqrt{T/m}$$

n, l, and T are measured variables. Determine m.

8. The velocity of outflow of an ideal fluid from a hole in the side of a tank is given by

$$v = \sqrt{2P/\rho}$$

v and P are measured variables. Determine ρ.

9. A conical pendulum has a period given by

$$T = 2\pi \sqrt{l \cos \alpha / g}$$

T and α are measured variables, l is fixed and known. Determine g.

10. The deflection of a cantilever beam follows

$$d = \frac{4Wl^3}{Yab^3}$$

d, W, and l are measured variables, a and b are fixed and known. Determine Y.

11. The capillary rise of a fluid in a tube is given by

$$h = \frac{R\sigma}{\rho g R}$$

h and R are measured variables, ρ and g are fixed and known. Determine σ.

12. The gas law for an ideal gas is

$$pv = RT$$

p and T are measured variables, v is fixed and known. Determine R.

13. The Doppler shift of frequency for a moving source is given by

$$f = f_0 \frac{v}{v - v_0}$$

f and v_0 are measured variables, f_0 is fixed and known. Determine v.

14. The linear expansion of a solid is described by

$$l = l_0(1 + \alpha \cdot \Delta t)$$

l and Δt are measured variables, l_0 is constant but unknown. Determine α.

15. The refraction equation is

$$\mu_1 \sin \theta_1 = \mu_2 \sin \theta_2$$

θ_1, θ_2 are measured variables, μ_1 fixed and known. Determine μ_2.

16. The thin lens (or mirror) equation can be written

$$\frac{1}{s} + \frac{1}{s'} = \frac{1}{f}$$

s, s' are measured variables. Determine f. There are two ways of plotting this function. Which is the better?

17. The resonant frequency of a parallel L-C circuit is given by

$$\omega = \frac{1}{\sqrt{LC}}$$

ω and C are measured variables. Determine L.

18. The force between electrostatic charges is described by

$$F = \frac{q_1 q_2}{4\pi\epsilon_0 r^2}$$

F and r are measured variables for fixed and known q_1, q_2. How do you check the inverse square law?

19. The force between currents is described by

$$F = \frac{\mu_0}{2\pi}\frac{i_1 i_2 L}{r}$$

F, i_1, i_2, and r are measured variables, μ_0 and L are constant. How do you check the form of the dependence?

20. The discharge of a capacitor is described by

$$Q = Q_0 e^{-t/RC}$$

Q and t are measured variables. R is fixed and known. Determine C.

21. The impedance of a series R-C circuit is

$$Z = \sqrt{R^2 + (1/\omega^2 C^2)}$$

Z and ω are measured variables. Determine R and C.

22. The relativistic variation of mass with velocity is

$$m = \frac{m_0}{\sqrt{1 - (v^2/c^2)}}$$

m and v are measured variables. Determine m_0 and c.

23. The wavelengths of the lines in the Balmer series of the hydrogen spectrum are given by

$$\frac{1}{\lambda} = R\left(\frac{1}{4} - \frac{1}{n^2}\right)$$

λ and n are measured variables. Determine R.

24. The thermionic current emitted from a heated filament is described by

$$J = AT^2 e^{-\varphi/kT}$$

J and T are measured variables, k constant and known. Determine A and φ.

25. The specific heat S of a solid is to be measured by the method of mixtures giving

$$m_s S(T_1 - T_2) = m_w(T_2 - T_3)$$

m_s = mass of solid—measured variable

T_1 = initial temperature of solid—fixed and known

T_2 = final temperature of mixture—measured variable

T_3 = initial temperature of water—fixed and known

m_w = water equivalent of calorimeter and contents—measured variable

Find S.

23. The thermal conductivity K of a solid is measured by measuring the temperature gradient along a bar while heat is flowing, according to the equation

$$Q = KA\frac{T_2 - T_1}{d}$$

The quantity of heat Q is determined by measuring the temperature rise of a stream of water circulating round the cooled end. Thus,

$$Q = m(T_4 - T_3)$$

where m is the quantity of water flowing per second and $T_4 - T_3$ is the temperature rise.

m, $T_4 - T_3$ and $T_2 - T_1$ are measured variables, d and A are fixed and known. Determine K.

27. A continuous flow calorimeter experiment is to measure the heat capacity of a liquid S. An amount of electrical energy VI is supplied and the resulting temperature change $T_2 - T_1$ of the fluid is measured. m is the mass of fluid flowing in unit time. Heat is lost from the sides of the tube constituting a systematic error. We can, however, presume that such heat loss is proportional to the difference between the mean temperature of the fluid $(T_1 + T_2)/2$ and that of the surroundings, T_0. The final heat balance equation could therefore be written

$$VI = JmS(T_2 - T_1) + \text{const}\left(\frac{T_1 + T_2}{2} - T_0\right)$$

V, I, m, T_2, and T_1 are all measured variables, J and T_0 are constant and known. Determine S.

6 Experiment Evaluation

6.1 The Aims of Experiment Evaluation

Once the measurements have been completed, the aim of the work is to determine the range of validity of the measurements, compute the answer and estimate its uncertainty. At this stage it is well to remember that the experimental results are precious. They may have been obtained from a massive experimental program involving many people and a lot of money. At any level they may well be irretrievable. It is necessary, therefore, to extract the maximum value from the results by squeezing every bit of information out of them. One must be objective too, because, if the outcome proves disappointing, it is necessary to state the result honestly and realistically, and obtain from it the necessary guidance for future work. It is a familiar misunderstanding amongst students that they are in the laboratory to reproduce the known values of their experimental qualities. If they then get 960 cm/sec^2 for g, this is different from the "right" answer, and so they are "wrong." The "error" is then probably blamed on the apparatus. This deplorably ungrateful attitude is described merely to point out its naivetë and wrongness. If the student had taken the trouble to estimate his uncertainty he would have been able to quote his answer as 960 cm/sec^2 $\pm$ 30 cm/sec^2 and so would have realized that he is actually right. If he is going to grumble about anything, let it be the $\pm$30 cm/sec^2, but he should not feel guilty about it, if the experiment with normal effort is not capable of a precision greater than 3 per cent. Part of this mistaken attitude arises from the fact that one normally meets the accepted values of the constants of physics in a textbook, and not in the laboratory. In the texts one usually finds a casual mention of the accepted value of a particular constant. The fact may not be made clear that this number is the outcome of many years of hard work by dedicated and brilliant men of science. To gain such insight the future experimenter should read the history of the measurements of such quantities as the mechanical equivalent of heat or the velocity of light. (See Reference 12 in the Bibliography.) One should not be too casual about numbers such as these, and should not hope to reproduce them in two hours of work in the elementary laboratory.

The point which we wish to emphasize is that the result of the experiment should be stated honestly and objectively as a best value, its uncertainty, and its range of validity. Certainly the experimenter should strive earnestly to maximize the yield of the experiment by making his best value as reliable as possible and his limits of uncertainty as narrow as the experiment will permit, but it is in all cases important to be realistic.

The evaluation of the experiment is in four parts, computation of the individual quantities, consideration of the validity of the results, computation of the answer, and estimation of the over-all precision of the experiment. These will be considered in turn.

6.2 Computation of Elementary Quantities

The first step in experiment evaluation is the calculation of the elementary quantities of which the final result is composed. For example the simple pendulum experiment will yield a set of values of T as dependent variable and l as independent variable. The present purpose is to compute the values of l and T, and their uncertainties, which will enable the subsequent graphical analysis to be accomplished. The treatment of the observations depends on whether the precision of observations is scale-limited or statistical.

Consider first the case of the scale-limited type. In the example above l may have been measured with a meter stick. Repetition may have shown that the reading, within 1 mm, did not show fluctuations. We therefore have a single value for l quoted to the nearest millimeter. This will consequently (see page 13) carry an automatic uncertainty of $\pm\frac{1}{2}$ mm meaning that we are "almost certain" that the real value lies within this range. The set of l values which constitutes the range covered by the experiment is therefore a set of lengths, quoted to the nearest millimeter and each with an absolute uncertainty of $\pm\frac{1}{2}$ mm leading to the appropriate value for the relative uncertainty. This will hold for all readings with scale-limited precision.

If the readings have an uncertainty which is statistical in nature, the necessity is to compute from the duplicated readings a best value and its uncertainty. For example, the simple pendulum experiment might reveal that the uncertainty in the timing measurement is sufficiently large that duplication of the measurement does give differing values. The experiment will then have been arranged to allow for sufficient duplication of T values for each value of l to give a precision which is estimated to be acceptable. The present task is to compute the precision which was actually achieved.

As was pointed out on page 39, the numbers of readings in the samples normally found in the work of the physics laboratory are too low to permit one to draw any conclusions regarding the actual frequency distribution of the universe from which they were drawn. Therefore, one tacitly assumes that the set of readings is a sample from a Gaussian universe and applies the results based on a Gaussian distribution given in Chapter 2. For a Gaussian distribution the best estimate of the true value X and the universe standard deviation is the sample mean $\bar{x}$ and the sample standard deviation s [using Equation (2.9)]. For a discussion of what we mean by the "best" value see page 133. If the mean of a set of quantities is required when the members of the set are, themselves, of unequal precision, a procedure known as "weighting" must be followed. This will control the contribution of each quantity to the final average in proportion to its precision. The weighting of observations will be described in Sec. 6.6. Given a set of repeated measurements, one therefore reduces it to the mean and the standard deviation or the standard deviation of the mean. At this point bear in mind the warnings about σ estimates from small samples and so make sure the computations are significant. In general, it is not worth while using a statistical approach with fewer than 10 observations and, for particular purposes, many more may be required. At this stage the measurement of every quantity in the experiment has been reduced to a best value and its uncertainty, and the way is clear to the graphical analysis of the experiment.

6.3 Graphs

Regardless of whether the graph is to be merely an illustration of the behavior of a physical system or whether it is to be the key to assessing the experiment and calculating the answer, the aim is to set out the results so as to display their characteristics as clearly as possible. This will involve appropriate choice of scale and other physical arrangements. Ensure that the graph paper is large enough. It is a waste of time to plot observations having a precision of $\frac{1}{5}$ per cent on a piece of graph paper 12 cm $\times$ 18 cm where a typical uncertainty is perhaps 2 per cent. As we shall see later, we want the uncertainty on the points to be clearly visible or else valuable information is lost, and so it is necessary to make sure the graph paper is big enough. Make the graph fill the available area. This can be done by choosing the scales so that the general course of the graph runs at about 45° to the axes and by suppressing the zero if necessary. If one is plotting the resistance of a copper wire as a function of temperature, and the values run from 57 to 62 ohms, start the resistance scale at 55 ohms and run it to 65. If the scale is started at zero, the graph will look like a flat roof over a sheet of empty graph paper and convey no information at all. The only exception to this suggestion arises sometimes when, for pictorial reasons, one wants to keep the scale of some effect clearly visible and we may want to show it in relation to some zero. However, for computational purposes, make the graph fill the paper.

It is necessary to indicate the uncertainties of the variables on the graph. This will enable the experimenter to judge how far the results show straight line behavior. Unless the range of possible variation is marked on each point, it is impossible to say whether any particular departure from the general trend is significant or not. Furthermore, the range of possibility for each point will determine the overall precision of the experiment as described on page 131. When the variables used are compound variables, the uncertainty in the compound quantity will have to be calculated for each point. The best way of marking the points on the graph is by a very fine dot at the center of bars indicating the possible range of variation. Make sure that the nature of this possible range of variation is clear, outer limits of possibility, $1s$ or $2s$ limits etc. When using compound variables it is very frequently useful to choose different symbols to distinguish the points arising from different values of a subsidiary variable. In this way a clue to the source of any discrepancy is provided (see Fig. 7.2) where the influence of differing pipe diameters on the discrepancy, the onset of turbulence, is thus clarified).

6.4 The Validity of the Experiment

If the experiment has been analyzed in terms of a straight line graph, the problem is to decide to what extent the measured points can be said to fall on a straight line. This decision will almost certainly be complicated by the inevitable scatter of the points and it is for this reason that the range of uncertainty of each point is plotted with it. The best way of judging a set of plotted points is to hold the graph paper at eye level and sight along the points. This makes the trend much more obvious than does the direct view. The clustering of the points about any linear region and any departures from linearity become quite clear. In this way a linear region of the curve can be identified and any discrepancies isolated. If no linear region is observed, contemplate the possibility that it lies below the lowest point, above the highest or in between two of the measured points, thus seeking guidance to future experimenting. Even if theory suggests that the graph should pass through the origin, do not take this into account when judging linearity unless the origin is actually a measured point. If the graph determined by the measured points fails to pass through the origin, this is precisely one of the discrepancies we wish to uncover.

Once a portion of the graph has been selected as a straight line, it is useful directly to draw in what is considered to be the best straight line. A method will be given later for the mathematical computation of this, but it is surprising how good visual judgment can be. A black thread stretched along the points, or a transparent ruler is good for drawing the best straight line. An opaque ruler is unsuitable. The final result will be as illustrated in Fig. 5.2, and the region of validity, in which the theory and experiment can be assumed to be compatible, has been determined. If the curve is completely linear as far as can be detected, the theory can be assumed to be adequate for calculation of the answer from all points. Note that this might not be the case if the experiment were to be repeated with increased precision. In this case the validity of the experiment would have to be re-checked.

6.5 The Calculation of the Answer and Its Uncertainty

If the preliminary analysis of the experiment has been satisfactorily accomplished, the final evaluation of the answer will present no problems. The answer, or some simple function of it, will be directly obtainable from the slope or intercept of the graph. The slope is, of course, not the geometrical slope of the line on the graph paper, for this depends on the scales chosen. The slope must be obtained by selecting two points on the graph, reading off the values of the variables for these points, and calculating the ratio of the difference in the value of the variables in the vertical direction to that in the horizontal direction. The dimensions of this slope will emerge automatically as the dimensions of the ratio of the two measured variables. In the case of the simple pendulum illustration, the dimensions will be that of l/T^2 or LT^{-2}. Check that this is appropriate to the answer required and make sure that the units are correct. The final step will probably be a simple calculation from the slope to the required answer. For the simple pendulum the slope was $g/4\pi^2$. The slope and π are known, hence g. In all cases of numerical computation, enormous simplification and reduced possibility of error will result from the use of a logarithmic mode of expressing very large or very small numbers. Thus one writes 176,000 as 1.76×10^5 and 0.00023 as 2.3×10^{-4}. In extended computation it is necessary to guard against accidental accumulation of rounding off errors. For this reason it is safer to work through a long calculation using more figures than are indicated to be significant by the precision of the original observations. Note that the normal slide rule does not give more than three significant figures, and that for only part of the scale. In extended calculations such rounding-off errors can easily accumulate to constitute an uncertainty greater than the experimental uncertainty. The use of a slide rule in such a case would be a very foolish waste of valuable information, and for all but the very simplest of calculations the use of a set of log tables is indicated.

If some deviation from linear behavior is found, it may suffice merely to have its presence detected and its influence on the answer eliminated. If, however, any speculation on its origin is required, the deviation from linearity must be treated as an observation of a new phenomenon. Perhaps someone can make a lucky guess about its origin, and one can always use the function-finding techniques to be described in Sec. 6.7.

The uncertainty in the answer is determined by the uncertainty in the slope from which it was determined. This uncertainty arises from the finite range of possible positions for each point given by its plotted uncertainty. This means that the points define a band of values, delineated by A_1A_2 and B_1B_2 in Fig. 6.1, instead of a unique line. The result of this is that a whole set of lines are contained within the limits of possibility. The central one of these will already have been chosen as the best line, CD, but limits exist, shown as A_1B_2 and A_2B_1, within which no slope is precluded by the observations. It must be admitted, therefore, that all slopes within these limits are possible, and the slopes of

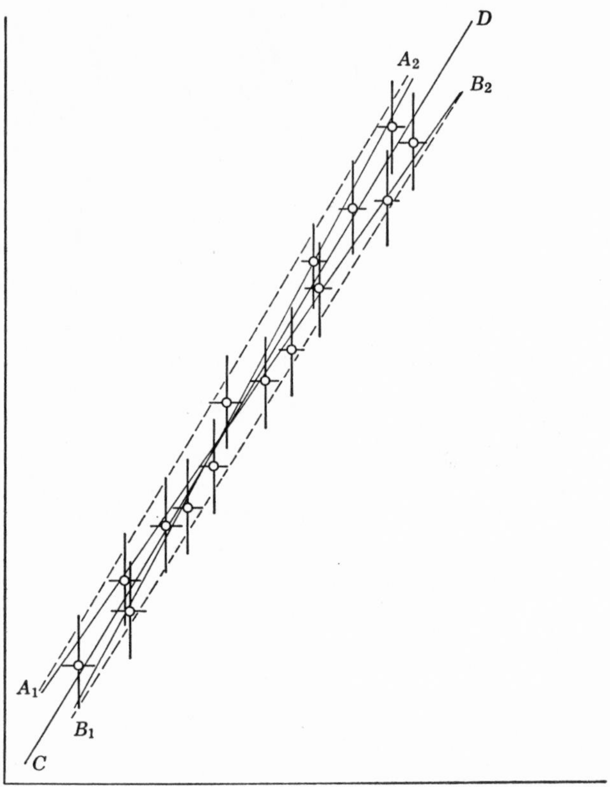

Fig. 6.1 The uncertainty in slope of the best straight line.

these limiting lines therefore represent limits of uncertainty for the slope of the best line. This uncertainty for the slope is thus obtained by calculating, in the same way as for the best line, the slopes of the limiting lines.

The uncertainty in the answer is then calculated from the theoretical expression for the slope, using the uncertainty in the slope and the uncertainties of any other measured quantities. It is necessary at this stage to recall clearly what kind of uncertainty has been marked on the graph. If the limits are outer limits of possibility (scale-limited or perhaps $2s$) then the limits on the slope will similarly be of this nature. If the points have been marked with $1s$ limits, the limiting slopes will be not quite outer limits of uncertainty, but will probably be safer than 68 per cent confidence since the limiting lines are drawn with a pessimistic bias.

There are really two cases to be considered. If the scatter encountered in the actual results is within the predicted range of uncertainty, then the use of the limiting lines gives rise to a fairly well-defined value for the uncertainty in slope. If, however, the scatter found is well outside the expected range of uncertainty (due to an unsuspected source of perturbation) then there is no unique setting for lines within which we are "almost certain" the answer lies. In such a case and in all precise work there is no substitute for the method of least squares to be described in the next section. In general, remember that experimental results are very precious quantities. If the experimenter is to justify his existence he must extract the maximum amount of information from them, either for his own answer or as a guide to future work.

6.6 The Principle of Least Squares

What is meant by the "best" value from a measurement process? It is merely begging the question to answer, "The one that stands the highest chance of coming close to the true value." Indeed there is no unique answer to our question but we can rephrase it as follows: If we have a set of observations x_1, x_2, x_3, . . . and we try to find best estimates of the values X and σ appropriate to the universe of readings we are, in effect taking a whole set of Gaussian curves with differing X and σ and trying them on our x_1, x_2, x_3, . . . x_n for fit. If we can find some process or criterion for our choice of the "best" X and σ whereby, on repeated sampling, i.e., by repeatedly taking batches $x_1 \ldots x_n$ of the readings, the values of our estimates of X and σ cluster most closely around some central value, and in an unbiassed fashion, then we can use this process to define a "best" value.

A similar consideration holds if we have a set of observations which we have plotted in the hope of obtaining a straight line. It will essentially show scatter as shown in Fig. 6.2. The question really is—does this set of observations provide support for any particular hypothetical variation and, if so, how much support? It is possible to draw a straight line through these observations but it is also possible, as shown in Fig. 6.2, to draw any number of other curves. What is to be the choice? The mathematical answer is, as stated above for repeated readings of a single quantity, that curve whose parameters show least variance on repeated sampling. By that we mean the following: If we consider the unperturbed (i.e., the ideal) variation to have the form

$$y = ax$$

we could use some criterion (as yet undefined) to compute a best estimate of a. If we take another sample of observations we shall obtain another best estimate. Repeated sampling gives a whole series of best estimates of a with a certain spread.

Suppose now we consider the unperturbed variation to be of the form

$$y = ax + bx^2$$

We could use the same criterion (still undefined) to obtain, on repeated sampling, a whole series of best estimates of a and b. The question now is—which postulated ideal variation gave the least spread in estimates of a? Whichever postulate it was, we can accept it as the better guess. However, who is to say that there is not a yet better guess, as yet unguessed? What then, is the justification for accepting any postulated ideal variation to describe a set of observations? The answer is almost invariably that someone has made a prior suggestion regarding the form of the function. If someone has taken a lot of trouble to make up a theory,

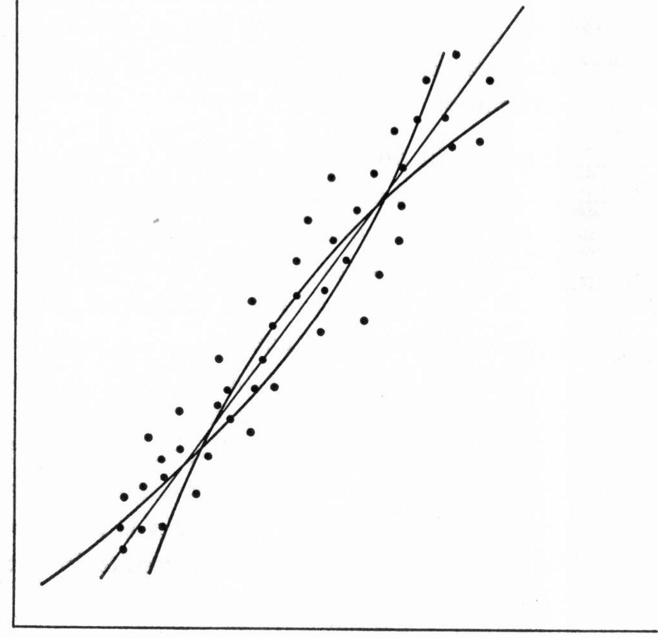

Fig. 6.2 The possibilities for a functional relation in observations showing statistical uncertainty.

the cause of science is best advanced by interpreting the results in terms of it. Whether it fits well or not the comparison will be the most useful way of progressing from the present state of knowledge to an improved state.

Granted, then, that a decision is made by the experimenter to interpret his observations in terms of some particular function, we must ask—what is the best way of determining the parameters of the function? We shall assume that the experimenter has been able to reduce his postulated function to straight line form

$$y = mx + b$$

and his problem will be solved if he can find some criterion for choosing the best estimate of the parameters m and b so that, on repeated sampling, he would find the least variance of his estimates.

This is really all one can say about the nature of a best value and, because of the vagueness, there has been much discussion over the choice for the criteria mentioned above. One criterion stands out through its simplicity and widespread use. This is the principle of least squares. It must be remembered however, that it is just a way of satisfying the requirements listed above for the choice of a best value, and is not a magical way of attaining absolute truth. It must be clearly borne in mind that, when a straight line fit is being adjusted by least squares, the answer for a slope or intercept has no greater validity than is possessed by the original decision to interpret the results in terms of a straight line.

The writer makes no apology for this wordy preamble to a very simple and commonly used technique. It is quite easy (and widely practiced) to use the least squares method without any idea of the significance of the procedure. It is the conviction of the writer that users should be aware not only of its powers but also of its limitations.

For the present purpose we shall accept the principle as an axiom. It is stated as follows:

> "The most probable value of an observed quantity is such that the sum of the squares of the deviations of the observations from this value is a minimum."

Let us consider the applications of the principle to two cases.

(a) *Repeated Observations of the Same Quantity*

Consider n observations $x_1, x_2, x_3, \ldots$ and call the most probable value X. Application of the least squares principle gives

$$\Sigma\,(x_i - X)^2 = \text{minimum}$$

Let $\bar{x}$ be the mean of the x_i. Then

$$\begin{aligned}\Sigma\,(x_i - X)^2 &= \Sigma\,[(x_i - \bar{x}) + (\bar{x} - X)]^2 \\ &= \Sigma\,[(x_i - \bar{x})^2 + (\bar{x} - X)^2 \\ &\qquad + 2(x_i - \bar{x})(\bar{x} - X)]\end{aligned}$$

or since $\Sigma\,(x_i - \bar{x}) = 0,$

$$\Sigma\,(x_i - X)^2 = \Sigma\,(x_i - \bar{x})^2 + (\bar{x} - X)^2$$

This last expression is clearly a minumum when $\bar{x} = X$, thus confirming that the use of the mean as the most probable value of a sample is consistent with the principle of least squares.

If the individual x_i do not all have the same precision, but have each a standard deviation, it can be shown that the principle of least squares gives for the best value

$$X = \frac{\Sigma\, x_i/s_i^2}{\Sigma\, 1/s_i^2} \tag{6.1}$$

This equation clearly biases the value X towards those x_i which have the smallest s_i, thus favoring the more precise measurements. This process is called "weighting" and X is a "weighted mean." It can be shown that the standard deviation of such a weighted mean is given by

$$\sqrt{\sum \frac{(x_i - \bar{x})^2}{s_i^2} \Big/ (n-1)\, \Sigma \frac{1}{s_i^2}} \tag{6.2}$$

(b) *Best Fit to Straight Lines*

Suppose that an experiment has yielded a set of n values of y as a function of x. Assume that, for the sake of simplicity, we can regard the values of x as exact, and that the values of y are subject to uncertainty.

When plotted, these readings appear as in Fig. 6.3. The task is to determine the constants of the straight line AB which best fits the observations. Let the equation of this line be $y = mx + b$ so that our objective is the best values of m and b.

The scatter of the points about the line is characterized by differences of the form

$$\delta y_i = y_i - (mx_i + b)$$

The criterion of least squares gives the constants m and b from the condition

$$\Sigma\, [y_i - (mx_i + b)]^2 = \text{minimum}$$

Write
$$\Sigma\, [y_i - (mx_i + b)]^2 = M$$

Then the condition for the minimum is

$$\frac{\partial M}{\partial m} = 0 \quad \text{and} \quad \frac{\partial M}{\partial b} = 0$$

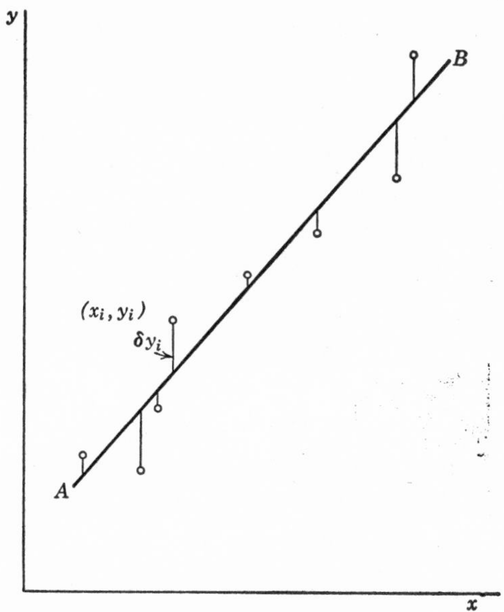

Fig. 6.3 The least squares fit to a straight line.

Elementary algebra then gives for the constants m and b

$$m = \frac{n \sum (x_i y_i) - \sum x_i \sum y_i}{n \sum x_i^2 - (\sum x_i)^2}$$
$$b = \frac{\sum x_i^2 \sum y_i - \sum x_i \sum (x_i y_i)}{n \sum x_i^2 - (\sum x_i)^2} \qquad (6.3)$$

This then defines the best straight line by a rather more acceptable criterion than visual judgment. Since it is statistically significant we can assign standard deviations to the estimates of m and b, so that their reliability has a definite meaning. They are expressed in terms of a quantity s_y which is the standard deviation of the δy values about the best line. This is given by

$$s_y = \sqrt{\frac{\sum (\delta y_i)^2}{n - 2}} \qquad (6.4)$$

and s_m and s_b are given by

$$s_m = s_y \sqrt{\frac{n}{n \sum x_i^2 - (\sum x_i)^2}}$$
$$s_b = s_y \sqrt{\frac{\sum x_i^2}{n \sum x_i^2 - (\sum x_i)^2}} \tag{6.5}$$

For a more complete, mathematically detailed treatment of the method of least squares the reader is referred to Appendix 2.

Functions other than straight lines can be fitted to observations by the method of least squares but Equations (6.3), (6.4), and (6.5) are, of course, not applicable. Recourse to first principles as described in Appendix 2 is necessary and the methods used will be found in Reference 2 in the Bibliography. If the uncertainties of the points are known to be different and values for the uncertainties are available, a weighting procedure can be followed to bias the m and b values in favor of the more precise points. A description of this weighted least squares method will be found in Reference 18 of the Bibliography.

6.7 Function Finding

It has been said before that, if the experimenter has any clue at all to the expected behavior of a system, he should use it in analyzing the results. Failure to do so would mean that there is some information in the results which is not being extracted. It may happen, however, that there is absolutely no information available to aid in the interpretation of a set of results. In such a case algebraic analysis of the results has a twofold aim. First, the establishment of some function which fits the observations adequately will be very useful to others who wish to use the results empirically. Although values can be read directly off a graph and this is very convenient, the representation of the results by an equation is much shorter and neater. Second, the discovery of some function which describes the results may (but only "may") be of assistance in the establishment of a model for the system. In any case it is worth trying.

The information available to the experimenter is a set of readings $y = f(x)$ which, when plotted simply in terms of the measured variables y vs. x (for lack of any better proposal), give a smooth curve of no readily identifiable shape. It is desired to obtain the form of the function $f(x)$ and this is easy in two particular cases, fortunately both of reasonably common occurrence in physical systems.

(a) *Power Law*

Suppose

$$y = x^a$$

where a is a constant.

Then

$$\log y = a \log x$$

Consequently a plot of $\log y$ vs. $\log x$ is a straight line of slope a. If a power law is suspected, therefore, plot the variables logarithmically, either on log – log graph paper, or on linear graph paper by obtaining the logs from tables. If a straight line is obtained, a power law dependence for the results can be claimed, with an index equal to the slope.

(b) *Exponential Law*

Many physical phenomena follow an exponential relation,

$$y = ae^{bx}$$

where a and b are constants. In this case

$$\log y = \log a + bx$$

and a plot of log y vs. x gives a straight line. If, therefore, this type of variation is suspected, either plot log y vs. x on linear graph paper or y vs. x on semi-log paper. Obtain the constant b from the slope and a from the intercept. It is necessary to be careful about intercept measurements, for if logs to the base 10 instead of to the base e are used, the origin is shifted.

If neither of these processes works, it is possible to express any set of readings as a power series of one variable as a function of the other using a difference table (see Reference 17 in the Bibliography). This polynomial representation may be useful for empirical work but it is still difficult to guess from it at the basic function which would be revealed by a theoretical treatment.

6.8 Over-all Precision of the Experiment

At the beginning of the experiment, the experimenter looked forward with a guess at the limits of uncertainty likely to be found. This was only an estimate for his own information. At the end of the experiment he should look back and, by a critical assessment of the results, evaluate the precision *actually achieved.* To be useful at all to other people, this figure must be realistic and honest, and it is worthwhile taking pains to make it reliable. On the other hand, there is no intention of persuading the student to immerse himself in a set of formal calculations, blindly following the mathematical requirements and finally quoting an uncertainty to three supposedly significant figures. Due to obscurity of the main issues it is quite likely to be significant to none. It is far better to make a realistic estimate guided by reason, than it is to lose sight of the point by overemphasis on the formalism.

The kind of uncertainty to be quoted does not matter *provided the quantity used is made clear.* Thus the experimenter may quote outer limits for the range of his answer (we presume he means something like 95–99 per cent confidence), or he can quote the standard deviation of his set of results or the standard deviation of the mean. It does not matter, but only so long as it is stated quite clearly what kind of quantity is being quoted. It is essential, of course, to ensure that all known sources of uncertainty are included in this evaluation of the precision. There is no point in claiming that potentials read off a 1 m slide wire potentiometer are precise to $\frac{1}{5}$ per cent just because the balance point could be read to within a millimeter. It could well be that lack of uniformity in the slide wire introduces uncertainties much greater than $\frac{1}{5}$ per cent. Similarly a thermometer could perhaps be read to $\frac{1}{10}$ deg but errors in calibration could be even $\frac{1}{2}$ deg. The lower uncertainty can, of course, be claimed as the uncertainty of reading the scale, but we are concerned at present with the *over-all* precision of the experiment.

It is assumed, then, that the readings at this stage have been pruned of all systematic deviations, and the remainder of the uncertainty calculation depends on circumstances.

(a) *Result Is the Mean of a Set of Readings*

The best quantity to quote is the standard deviation of the mean, since this has an immediate numerical significance. Sometimes the standard deviation itself is quoted. In all cases it is essential to quote the number of readings so that the reliability of the σ estimate can be judged.

(b) *Result Is the Consequence of a Single Calculation*

In the undesirable event that no graphical analysis has been possible and the result is obtained algebraically from a number of measured quantities, use the methods of Chapter 3 to calculate outer limits for the uncertainty, or the standard deviation.

(c) *Result Is Obtained Graphically*

If the straight line has been established by a least squares method, the uncertainties in the constants m and b are obtained directly. Note that these uncertainties have the advantage that they have been obtained from the actual scatter of the points, regardless of their estimated uncertainties. (This does not mean, however, that if one intends to make a least squares fit to a straight line, one should not bother to plot in the uncertainties, or even to draw a graph at all. The graph, with the uncertainties on the points, is still needed to judge the range of validity of the experiment, and thus to detect possible systematic errors. A least squares fit should be attempted only after one has, by visual inspection of the graph, selected the range which is to be considered linear). If the straight line has been drawn by eye, the lines at the limits of possibility will give the possible range of slope and intercept. This uncertainty in slope will then probably have to be combined with the uncertainties of some other quantities before the final uncertainty of the answer can be stated.

In all cases, it probably does not matter too much what kind of uncertainty is quoted, so long as one is quoted, and the nature of the quoted value is made clear.

When working through lengthy uncertainty calculations the arithmetic may be simplified by dropping insignificant uncertainty contributions. There is no point in adding a 0.01 per cent contribution to one of 5 per cent. In the final statement of uncertainty it is not commonly valid to quote uncertainties to any more than two significant figures. Only work of real statistical significance would justify more.

Once the over-all uncertainty of the final answer has been obtained, the question of the number of significant figures to be retained in the answer can be considered. There is no unique answer to this question but, in general, one should not keep figures after the first uncertain figure. For example, 5.4387 ± 0.1 is 5.4 because, if the 4 is uncertain, the 387 are much more so. However, if the uncertainty is known more precisely, it might be justifiable to keep one more figure. Thus, if the uncertainty above were known to be 0.15, it might be possible to quote 5.44 ± 0.15. When quoting observations with scale-limited uncertainty one cannot use the suggestion of keeping one uncertain figure. A meter stick reading is quoted as 43.2 ± 0.05, not as 43.20 ± 0.05. This is because the statement of the measurement is that the value lies somewhere between 43.15 and 43.25, and 43.2 is just the name which is given to this interval.

If a measurement is quoted with a percentage precision, the number of significant figures is automatically implied. For example, a measurement might be quoted as 527.64182 ± 1 per cent. This means that the absolute uncertainty is around 5.276. However the precision itself is quoted to only one significant figure (1 per cent, not 1.00 per cent) so that we are not justified in using more than one significant figure in the absolute uncertainty. We shall call the absolute uncertainty 5 and this implies that, if the 7 in the original number is uncertain by 5, the .64182 has no meaning. The measurement could then be quoted as 528 ± 5. If a set of readings has yielded a mean as the answer, the number of significant figures in the mean will be governed by the standard deviation of the mean, and the number of significant figures in the standard deviation will be governed, in turn, by the standard deviation of the standard deviation.

The experimenter should make sure that he quotes his answer and his uncertainty in such a way that they are consistent, i.e., neither as 16.2485 ± 0.5 nor as 4.3 ± 0.0002.

PROBLEMS

1. A value is quoted as 14.248 ± 0.1. State it with the appropriate number of significant figures.

2. If the value is quoted as 14.248 ± 0.15, how is it quoted with the appropriate number of significant figures?

3. A value is quoted as 6.74914 ± $\frac{1}{2}$ per cent. State it with an absolute uncertainty, both with the appropriate number of significant figures.

4. State the results of the calculations in the problem at the end of Chapter 2 as a best value and an uncertainty, both with the appropriate number of significant figures.

5. An experiment was done to measure the impedance of a series

R-L circuit. The impedance Z is given as a function of the resistance R, the frequency of the source f and the inductance L by

$$Z^2 = R^2 + 4\pi^2 f^2 L^2$$

The experiment was done by measuring Z as a function of f with the intention of plotting Z^2 vertically and f^2 horizontally to obtain L from the slope and R from the intercept. The observations obtained are given in the table.

f c. sec^{-1}	Z ohms	f^2	$f.(AUf)$	$AUf^2 = 2f(AUf)$	Z^2	$Z.(AUZ)$	$AUZ^2 = 2Z(AUZ)$
123 ± 4	7.4 ± 0.2						
158	8.4						
194	9.1						
200	9.6						
229	10.3						
245	10.5						
269	11.4						
292	11.9						
296	12.2						

The uncertainties given in the first line refer to all the readings in each column.

(a) Plot these readings in the appropriate fashion and mark the uncertainties on the points. Suggested table headings to expedite the calculations are given above.

(b) Check to see if the observations can be interpreted in terms of a straight line for any part of the range or all of it.

(c) Obtain the slope of the best line.

(d) Calculate the best value for L.

(e) Obtain the slopes of the lines at the outer limits of possibility and so state the range of uncertainty for the slope.

(f) Calculate the absolute uncertainty in the measurement of L.

(g) Obtain the best value of R from the intercept.

(h) Obtain the uncertainty for the R value.

(i) State the complete result for the experiment with the appropriate number of significant figures in each quantity.

6. Ten different observers report on the intensity of a lamp measured with a comparison photometer. Their results (in arbitrary units) are as follows:

Observer	Mean	Standard deviation of mean
1	17.3	2.1
2	18.4	1.9
3	17.1	2.5
4	16.6	2.8
5	19.1	3.2
6	17.4	1.2
7	18.5	1.8
8	14.3	4.5
9	16.8	2.3
10	17.4	1.6

What is the result of the experiment and what is its standard deviation?

7. An experiment has been carried out to investigate the temperature dependence of the resistance of a copper wire. The ideal variation can be represented by

$$R = R_0(1 + \alpha T)$$

where R is the resistance at temperature T°C, R is the resistance at 0°C and α is the temperature coefficient of resistance. The following observations of R and T were obtained:

Calculation of best values of R_0 and α				Calculation of standard deviations using calculated best values of R_0 and α			
x	y	xy	x^2				
T°C	R ohms	TR	T^2	$R_0\alpha T$	Calculated ideal value of R $(= R_0 + R_0\alpha T)$	δR (obs. R − ideal R)	$(\delta R)^2$
10	12.3						
20	12.9						
30	13.6						
40	13.8						
50	14.5						
60	15.1						
70	15.2						
80	15.9						
Σx and $(\Sigma x)^2$	Σy	$\Sigma (xy)$	$\Sigma (x^2)$				$\Sigma (\delta y)^2$ and so, s_y

(a) Using the method of least squares (suggested headings under which the calculations for this can be carried out are given in the first part of the table) obtain the best value for the slope and for the intercept.

(b) Hence obtain the best value for α.

(c) Evaluate the standard deviation for the slope and for the intercept (suggested headings for this part of the calculation are given in the second part of the table).

(d) Hence evaluate the standard deviation of α.

(e) State the final result of the experiment with the appropriate number of significant figures.

8. The following measurements were made in the investigation of new phenomena. In each case identify the function and evaluate its constants.

(a)

v	i
0.1	0.61
0.2	0.75
0.3	0.91
0.4	1.11
0.5	1.36
0.6	1.66
0.7	2.03
0.8	2.48
0.9	3.03

(b)

x	y
2	3.2
4	16.7
6	44.2
8	88.2
10	150.7
12	233.5
14	337.9
16	464.5
18	618.0

(c)

T	f
100	0.161
150	0.546
200	0.995
250	1.438
300	1.829
350	2.191
400	2.500
450	2.755
500	2.981

(This last one is a little less obvious)

7 The Scientific Report

7.1 Aims in Report Writing

The value of skill in descriptive writing to any scientist or engineer can hardly be overestimated. The written report, whether privately circulated in a small organization, or published in an international journal, is the standard mode of communication in all forms of technological or scientific work. It is very frequently the only contact between the experimenter and those potentially interested in his work and so it must be able to stand by itself as the record of the experiment and its interpretation. The aim in the report, therefore, has to be to convey to the reader as clearly, concisely and convincingly as possible the description of the experimenter's work. The development of a fluent, elegant style of writing should not be regarded as a waste of time even for the most technical-minded of engineers. Irritation on the part of the reader of a report which is caused by obscurity, repetition or failure to create a logical sequence can depreciate even the best experimental work. So often, on reading a poor and obscurely written report, the reader finally puzzles out what the writer meant to say and says to himself in exasperation, "Why didn't he just say so?"

Try to avoid inducing this feeling in the readers of your reports.

The circumstances of report writing can vary so much that it is, unfortunately, impossible to lay down specific rules for good reports. The range of topic and purpose is far too wide, and it is the student's responsibility to acquire his own style for his own purpose. This can happen only through practice and correction, and such practice is the main purpose in writing laboratory reports. They should therefore be treated, not so much as mere assignments to be completed, as opportunities for the practice of good descriptive writing. This practice is necessary since the technique of clear explanation (either written or spoken) does not necessarily come naturally. For this purpose there is no substitute for personal application, but one very useful method of absorbing some of the atmosphere of the literature in a particular field is to look at the pages of a reputable journal. This will show the student, at any stage, the standard at which he is aiming.

7.2 General Principles

This section contains some hints which may help the student to become fluent in scientific writing. It is intended to be very general but is illustrated specifically with reference to the physics laboratory.

The most important single point is clarity. The aim of the report is to convey information and it is a failure unless it does so. At all stages of the writing, pause to ask, "What is the point I wish to make here?" Once this is settled, put it down in a simple and straightforward fashion so that it emerges clearly. Obscurity can arise as easily from too many words as from too few, so examine every statement for information density. In almost all cases the most concise way of saying something is the best, and one should discard everything which does not tell the reader something significant. The report should be a logical sequence, one section leading into another without abrupt changes in the train of thought. The reader should be led gently through the argument by a series of smooth, logical steps, all the way through the report from the first statement of the topic of the work to the final conclusions. One very useful method of keeping the main line of argument clear is the use of appendices. Any detail of theory, method or discussion or any lengthy table of results which, although judged worthy of inclusion, would obscure the main line of the argument should be relegated to an appendix. To do this, divide material relating to the experiment into three categories; descriptive, reference, and personal. Put the first in the text of the report, put the second in the appendices and do not put the third in at all. People are not interested in the trials and tribulations of your experiment unless you have some really specific point which will help them to avoid difficulties. The place for your record of trials and tribulations is your own note book. Write the report in the past tense and avoid first person constructions and colloquial expressions. Pay attention to the format of the report, taking care to give the various sections a clear heading which stands out. The reader should not have to hunt around in a featureless maze of text and numbers. In a long, complicated report a table of contents at the beginning is very helpful.

The general rule, one which will be of great assistance in writing good reports is—think of your reader all the time. Ask yourself what he wants to know, and tell it to him. Your reader constitutes the purpose in writing the report and he is one of the most important features of it.

It is not intended to suggest a rigid format for the report since circumstances vary so much, but some features of the report are of fairly general importance.

7.3 Introductory Material

The writer of a report must start with the assumption that the reader, although intelligent and well-informed, knows nothing about the work described in the report. The function of the introductory part of the report is, therefore, to bring the reader to a point at which a detailed description of the procedure and results makes sense.

The normal components of introductory material are:

(a) *Title*
(b) *Topic*
(c) *Background Material*
(d) *Application of the Background Material to the Particular Experiment*
(e) *Specific Statement of Intention*

These headings, (a) to (e), are not intended to be taken as suggestions for actual headings in a report. They are merely the items which constitute the analysis of a report and suggestions regarding headings for use in actual reports will be given later. The items above will now be considered in turn.

(a) *Title*

This should be short and to the point. Its purpose is only to identify the work, not describe it.

(b) *Topic*

The first thing the reader wants to know is the subject to be considered in the report. He has already had a hint of this in the title, but he is not ready yet for detail. Think of the most general statement about the experiment you can make and start with that. For example, an experiment on e/m for electrons could start with the statement, "The ratio of charge to mass for electrons can be obtained from the trajectory of an electron beam in a combination of electric and magnetic fields." This introduces the reader to the topic of the work generally, and this kind of topic statement will be found at the beginning of almost every good report. Make sure this topic statement is clear, because if the reader misses the point right at the beginning, the rest of the report may be of very little value.

(c) *Background Material*

The reader must be brought up-to-date in the field of the experiment, and must be provided with enough information to make the rest of the report intelligible. This background material may be either historical or theoretical. It is always a difficult point to know how much of this to include, since one does not know how much the reader already knows. A good guide is to put in the minimum that will render the rest of the report understandable. Unless you are writing a review paper for the purpose of expounding the work of other people, quote the results of established theory only, giving a reference where the whole treatment may be found.

Make sure, however, that any limitations on the result are stated if they are important in the experiment. Thus one would write, "It can be shown (Reference 1) that the period T of a simple pendulum of length l can, if the amplitude is sufficiently small, be written etc., etc." If it is considered that some detailed piece of theory is necessary, consider the possibility of relegating it to an appendix to keep the logical thread of the development clear.

(d) *Application to the Particular Experiment*

The reader must be shown how the background material or theory is relevant to the present work. In many cases the standard theory must be revised or extended. Since such application is not standard and is probably the work of the experimenter himself, it should be given at adequate length to enable its validity to be judged. This section will probably contain the analysis of the experiment in linear form, and will thus reveal how the validity of the experiment is to be checked and the result computed. This section may also contain a reference to any special measuring technique which may be necessary.

(e) *Specific Statement of Intention*

It is a great help to the reader if the position is summarized at this stage to say fairly precisely what is intended in the experiment, e.g., "Thus, if the variation of x and y for various values of z is measured, a straight line graph of xz^2 vs. y should be obtained and this will give a from the slope and b from the intercept on the horizontal axis."

The description above of introduction material may be subdivided to suit the particular circumstances. In an elementary experiment it need occupy no more than a single paragraph and be labelled simply, "Introduction." In a more complicated experiment, more extended discussion of some particular topic may be necessary, thus justifying other headings such as "Theory" or "Method." The actual mode of subdivision is probably unimportant (provided it is clear).

It will be noticed that an important concept, the aim of the experiment, has not yet been discussed. This is again because it is undesirable to warp the report to fit too rigid a format, and the place where the aim of the experiment may be stated varies very much according to circumstances. If the subject of the experiment is well known and well established, the aim makes a convenient topic statement, e.g., "The aim of the experiment was to measure *n* by the method of Jones." On the other hand, if a relatively obscure quantity is being measured, it might not be possible to state the aim until after a considerable amount of definition and theory had been given. It probably does not matter too much, provided only that the aim is clearly stated somewhere and probably as early as is convenient.

At this stage the reader has been told the topic of the experiment, given any necessary background, shown how this is applied to the particular experiment and finally how the whole scheme is used to obtain the required answer. This leaves the reader ready to be told how the experiment was actually done.

7.4 Procedure

A good plan to follow in the "Procedure" section is:

(a) *A Statement of the General Outline of the Performance of the Experiment:*

This should be stated as concisely as possible so that the reader can see immediately the general course of the experiment. If the experiment consisted of the measurement of the variation of y with x over the range a to b, say just that.

(b) *Detailed Description of Specific Measurement Methods:*

It is essential to ensure that the reader is told the method of measurement of every factor contributing to the final result, but this description should not degenerate into a list of experimental trivia. Insert only enough detail so that the reader can judge validity of the procedure. An exception to this advice is encountered in papers describing some new experimental procedure which is described so that other people can duplicate the method. Here, clearly, the description must be complete, but this is a special case and commonly a much more general description of procedure can be used. Well known, standard methods of measurement need be identified by name only, and only in special cases should they be described in detail.

The description of measurement method will, of course, involve the description of any necessary apparatus but make sure that the principle of the method is first described clearly and prominently. Do not ask your reader to follow through a long, detailed, step-by-step description of some complicated piece of apparatus only to have him ask himself at the end, "Yes, but what is it measuring and how?"

(c) *Any Additional Details of Care or Precaution Which Proved Necessary.*

There should be a diagram, with a figure number, of even the simplest apparatus, but it is rarely worth while making diagrams into works of art. They exist merely to aid the reader and should be as simple as possible. They should, however, be neat: it is not too much to expect the use of a ruler. Notations on diagrams form a very neat and useful way of providing reference-type information without cluttering the actual text. This is particularly true of circuit diagrams, on which the values of resistors, capacitors and other components, the ranges of meters, ratings of power supplies, etc. can be marked. If there is no room on the diagram itself, mark the components with a simple symbol, R_1, C_2, etc., and make a little table beside the diagram with the actual values on it, as suggested on Fig. 7.1.

If the main topic of the experiment is a study of the properties of a sample of material, the procedure section must

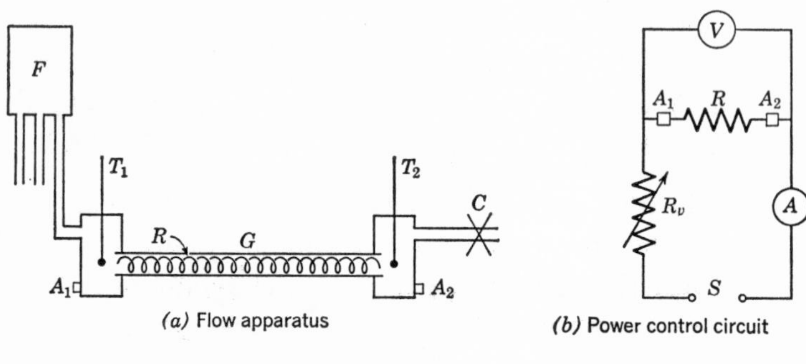

Fig. 7.1 A completed apparatus diagram.

contain an account of the nature of the sample, its origin, physical form, mode of preparation, etc.

If for any purpose, any particular section has to be treated in detail, e.g., method of making a particular measurement, preparation of samples, etc., make it into a section on its own, with its own heading. If this obscures the main line of the argument, put the material in the appendix.

7.5 Results

The main point here, again, is clarity. From the preceding sections the reader has been led to expect a set of values for the variation of y with x and measurements of a, b, and c. When he turns to the results page he wants to see just that. He should not have to work his way through a page, solidly filled with numbers and devoid of description and notations. If the principal result of the experiment is a set of values of y and x, put them in a big clear table, standing out by themselves and headed clearly "Variation of y with x." All other tables should also be clearly labelled. Give each table a number so that it is easier to refer to it in the text. Any subsidiary measurements should similarly stand out by themselves, clearly labelled and prominently displayed. Note that there is a difference between the table of results constructed by the experimenter in his laboratory note book and that required for the report. The former was laid out to expedite calculation and included a lot of information with which the reader is not concerned and whose presence would merely irritate him. In a column of results the reader wants results and only that. All intermediate arithmetic should be eliminated and values should have their uncertainties marked right beside them and not in a separate column.

Every measured value should have its uncertainty indicated. If the entries in a whole column of readings have the same uncertainty, this can be quoted just once at the top. Otherwise, every reading should be accompanied by an uncertainty, and the nature of the uncertainty (outside limits, standard deviation, etc.) clearly stated. Original readings should always be included, or else the mode of computation should be made so clear that the original readings can be recovered. If this is going to make the Results section too long and bulky, the lengthy portions can be reserved for an appendix.

Do not clutter up the Results section with calculations. Keep all the work on derived quantities (e.g., slopes) separate. Do not include trivial algebra and ensure only that sufficient detail is displayed to make the mode of calculation obvious. Any calculations which are needed but are lengthy should, again, be put in an appendix. At the end of the process make the final answer and its uncertainty really prominent.

7.6 Graphs

Graphs should be as clear and informative as possible. It is quite likely that the reader's judgment of the work will be based more on the graphical presentation of the results than anything else. The graph that was drawn during the evaluation of the experiment was for purposes of computation, and the emphasis was on fineness of drawing and choice of format to permit the maximum extraction of information from the results. The graph in the report is a copy of this one, but its role is more that of illustration, and attention should therefore be directed to clarity and completeness, as suggested in Fig. 7.2.

Choose suitable scales so that the results fill the available space to a reasonable extent. Label the axes clearly and include the units of measurement. Put a clear heading on

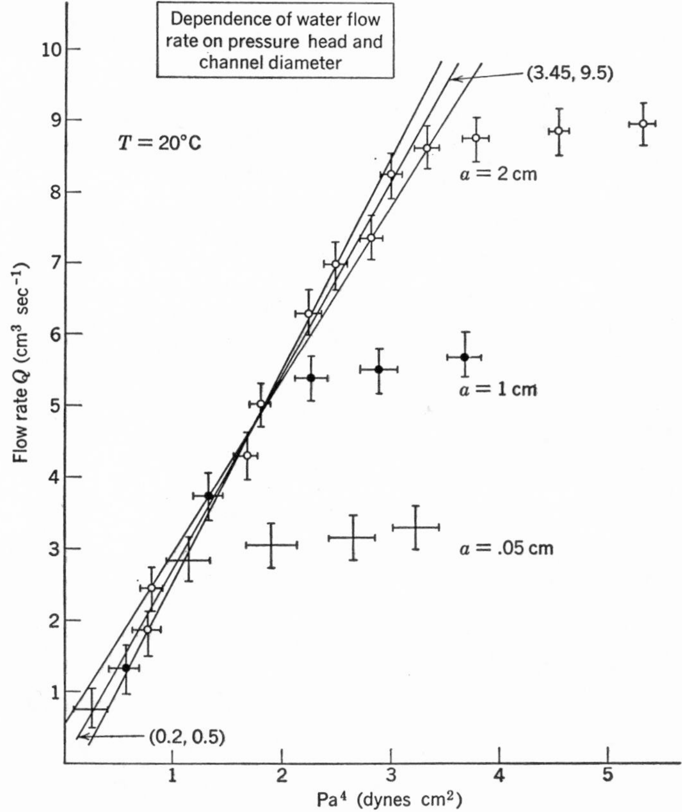

Fig. 7.2 A completed graph.

the graph so that there is no doubt about its identification. Indicate the range of uncertainty of each point. When a slope has been obtained from a graph, do not indicate this by a drawn-in triangle because this can obscure the position of the point from which the slope is to be calculated. Instead, the chosen point should be indicated in some way without interfering with the line, and the coordinates of the point written nearby. No other writing on the graph should be permitted since a tendency to fill up blank spaces with calculations, etc. merely obscures the message of the graph. Graphs should be drawn in ink. All illustrations in a report, apparatus diagrams, circuits, graphs, etc. should be given a figure number to facilitate reference in the text.

7.7 Discussion

A suggested list of topics is as follows:

(a) *A Summary of the Experiment and Conclusion:*

This should be inserted even if the experiment is a simple measurement: "n has been measured by the method of Jones and found to be 52.3 ± 0.5."

(b) *Comparison with Expectation:*

The experiment was very likely to have started with a piece of theory. As was stated on page 102, this was a prediction, on the validity of which the successful measurement of the unknown was based. In the successful performance of the experiment this validity was investigated. The extent to which it was confirmed should now be stated as a reminder to the reader and care should be taken to point out how any possible systematic error from any loss of validity (such as the onset of turbulence in fluid flow)

8.2 Experiment Technique

The range of choice of technique in the average student laboratory is rather limited. However, within the limits of this restriction, the student should learn to take responsibility himself for his choice of measurement conditions. He is going to have to accept this responsibility later on in professional life and first year university is none too soon to start training for it. He must therefore learn to decide to use a micrometer rather than a meter stick to measure a wire diameter, the range over which readings are desirable, and how many readings are necessary, etc. Even safety factors to a certain extent, have to be left to the student, so consider matters of personal hazard or apparatus limitations, and do not assume merely that someone else has already taken care of these things. This has all been stressed in the sections on experiment design, but the necessity for the student to plan his own experiment is also found at the level of setting up. The general advice is to think continuously about what one is doing and make every action the consequence of a considered experimental necessity. For example, in electrical work decide in advance the value and power rating of resistor requirements, and use these values instead of wiring in the nearest rheostat because it happens to have been left beside the apparatus. A large fraction of the sales of General Radio potentiometers must arise from failure to use a little preliminary thought.

There are two phases in the actual performance of the experiment; choice of apparatus and its actual manipulation.

(a) *Choice of Apparatus*

This choice is naturally limited in elementary laboratories, and any such restrictions will already have been incorporated in the measurement program. However, if choice is available, the particular item will be selected on the basis, once more, of the measurement program. This is the point at which to watch again for ratings. It would be foolish to write out an experiment plan calling for currents up to 10 amp. when the only available resistor is clearly marked with a maximum current of 5 amp. Always bear such ratings in mind, look for them on equipment, and plan the experiment accordingly. If there is no rating marked on a particular piece of equipment, find its instruction book or its entry in a catalogue and look there.

It is almost always best to tackle experimental problems (like all problems) in pieces. Consider each independent section of the experiment separately and thus simplify the work. Choose the experimental conditions to have as many built-in checks as possible (see page 113).

(b) *Performance of Experiment*

Make sure that your experimental arrangement is orderly and, especially, convenient. Good, thoughtful experimenting cannot be carried out if some delicate adjustment has to be made by stretching to the back of the bench over exposed terminals at 500 v while watching a meter at the front. It even helps merely to have it look nice. In general, use a thoughtful approach throughout. Make sure that you are in charge of the experiment, not the other way around. When setting up apparatus, treat the experiment in sections. Set up each section and ensure that it works correctly before starting on the next bit. If one assembles a huge, complicated piece of apparatus without such preliminary piece-by-piece checking, the probability that it all works the first time is close to zero, and the problems of identifying the trouble, or troubles, may be insurmountable. When constructing an electrical circuit always work from a carefully drawn circuit diagram. When starting up apparatus always start with the least sensitive condition, i.e. the maximum ranges on meters, maximum resistance in rheostats, minimum settings on potential dividers, etc.

Always check the zero on every instrument. Check, too, the calibration if at all possible, because errors in calibration constitute a type of systematic error which it is not normally possible to detect using the internal consistency methods of Chapter 5. In precise work it is not normally adequate to check instruments at one point only, and so they should be tested at all points on the range in use. When this is done a graph can be drawn giving either the corrected reading at each point on the scale, or else the error correction to be applied to each scale reading. Such a curve is called a *calibration curve* and is usually supplied with instruments designed for precision measurements.

The instruments most prone to errors of calibration are frequently the easiest to check, clocks and watches, thermometers, etc., and there is rarely any excuse for ignoring this. Electrical meters may be more of a problem if standards are not available. It is quite surprising how inaccurate cheap meters can be, and no measurement in which a precision of better than perhaps 5 per cent is desired should be based on them. In the case of a multi-range meter for which a detailed calibration is not possible, try to restrict the readings to one range of the instrument. This will not eliminate calibration errors but will at least ensure that one has only one potential calibration error to consider.

Calibration errors arise, too, from backlash in mechanical instruments. If the thread is loose on a micrometer eyepiece, a finite rotation of the screw (and consequently a finite reading on the scale) is required before the eyepiece cross wire starts to move. Errors due to backlash in such mechanical devices can be eliminated by taking all the readings going the same way, and care should be taken to ensure that this be done.

One practical point associated with the reading of instrument scales is the phenomenon of parallax. If the position of some index (e.g., the top of a barometer mercury column or an electrical meter needle) must be read on some scale which does not lie in the same plane as the index, the indicated value will vary with the position of the observer's eye. This is illustrated in Fig. 8.1 from which it can be seen that the error increases as the distance between the scale and the index increases. This source of uncertainty can be minimized in many ways. If the observer has control over the distance between the measured point and the scale, he

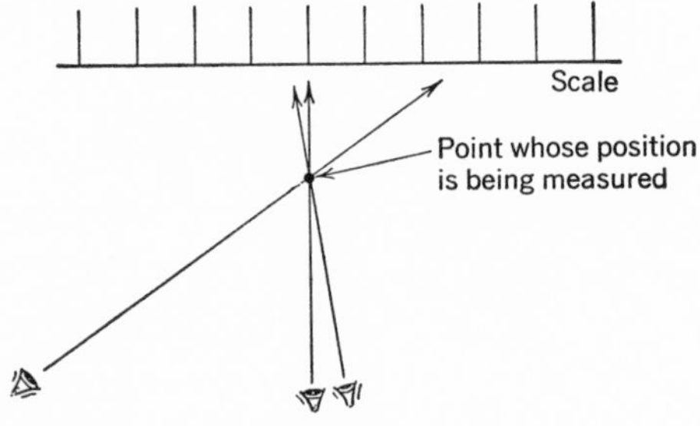

Fig. 8.1 Parallax.

should use it to bring them as close together as possible e.g., a meter stick should be used standing on its edge so that the graduations are actually in contact with the object to be measured. Frequently, a barometer will have the sliding part of its scale double, one part in front of and the other behind the mercury column. Therefore, one lines up the two parts of the scale with the top of the mercury column and thus eliminates angular effects. Good quality electric meters have a mirror in the plane of the scale so that if one lines up the pointer with its reflection a reading is ensured with the line of sight perpendicular to the scale.

During the actual taking of readings it is usually very helpful to plot a rough graph of the variables. This will provide a running check on the performance of the experiment and may suggest changes in the measurement program to maximize the yield of the experiment.

Appendix

The following sections are intended to give a more detailed mathematical treatment of some of the topics introduced in the main body of the text. Such a treatment was unnecessary at that point and would merely have served to interrupt the progress of the development. The reader who wishes a justification for many of the equations used will find it in these appendices. Any topic which is not mentioned cannot be treated adequately here and reference must be made to the more advanced texts.

1 The Statistics of the Gaussian or Normal Distribution

A1.1 The Equation of the Normal Distribution Curve

Consider that a quantity whose true value is X is subject to random uncertainty. Consider that the uncertainty arises from a number, $2n$, of errors each of magnitude E and equally likely to be positive or negative. The measured value x can then range all the way between $X - 2nE$ if all the errors should happen to have the same sign in the negative direction to $X + 2nE$ if the same thing happened positively. We require the distribution curve of a very large number of readings each subject to this uncertainty so that the quantity we must calculate is the probability of encountering a particular error R within this range of $\pm 2nE$. This probability is governed by the number of ways in which a particular error can be generated. For example, an error of the total value $2nE$ can be generated in only one way, all the elementary errors must have happened to have the same sign simultaneously. An error of magnitude $(2n - 2)E$, on the other hand, can occur in many ways. If any one of the elementary errors had been negative the total error would have added up to $(2n - 2)E$ and this situation can arise in $2n$ different ways. An error of $(2n - 2)E$ is, therefore, $2n$ times as likely as one of $2nE$. A situation in which two of the elementary errors had negative signs can correspondingly be generated in many more ways than the previous one and so on. This argument can be generalized, therefore, by using the number of ways in which a specific error R can be generated as a measure of the probability of the occurrence of the error R, and consequently as a measure of its frequency in a universe of observations.

Consider a total error R of magnitude $2rE$ (where $r < n$). This must be the result of some combination of errors of which $(n + r)$ are positive and $(n - r)$ are negative. The number of ways in which this can happen can be calculated as follows: The number of ways of selecting any particular arrangement of $2n$ items is $(2n)!$. However, not all of these arrangements are different for our purpose since we do not care if there is any internal rearrangement between the errors in the positive group. We must divide the total number of arrangements, therefore, by the number of these insignificant rearrangements, i.e., by $(n + r)!$. Similarly we must divide by the number of internal rearrangements possible in the negative group, i.e., by $(n - r)!$. The total number of arrangements is, therefore,

$$\frac{(2r)!}{(n + r)!(n - r)!}$$

This is not yet strictly a probability although it is a measure of the likelihood of finding such a total error. The probability itself will be obtained by multiplying the above number by the probability of this combination of $(n + r)$

positive and $(n - r)$ negative choices. Since the probability of each choice is $\frac{1}{2}$ the required multiplier is

$$\tfrac{1}{2}^{(n+r)} \; \tfrac{1}{2}^{(n-r)}$$

The final result for the probability of the error R is then

$$\frac{(2n)!}{(n+r)!(n-r)!}\left(\frac{1}{2}\right)^{(n+r)}\left(\frac{1}{2}\right)^{(n-r)} \tag{a1}$$

Our problem is now to evaluate this as a function of the variable r. This is done subject to the condition that m is very large, in fact tending to infinity.

The evaluation requires two auxiliary results.

(i) The first auxiliary result:

$$n! \approx \sqrt{2\pi n}\, e^{-n} n^n$$

This is known as *Stirling's theorem.* Although its full derivation is beyond our scope its plausibility can be indicated as follows:

$$\begin{aligned} \int_1^n \log x \, dx &= [x \log x - x]_1^n \\ &= n \log n - n + 1 \end{aligned}$$

The graph of $\log x$ vs. x is shown in Fig. A1.1. Clearly the integral above can be approximated by the sum

$$\log 1 + \log 2 + \log 3 + \cdots + \log n$$

which is $\log\,(1 \times 2 \times 3 \times \cdots \times n)$ or $\log n!$. We can, therefore, write approximately if n is large

$$\log n! = n \log n - n$$

i.e.,

$$n! = e^{-n} n^n$$

This is an approximation to the formula given above and the reader is referred to Reference 8 in the Bibliography for a full derivation.

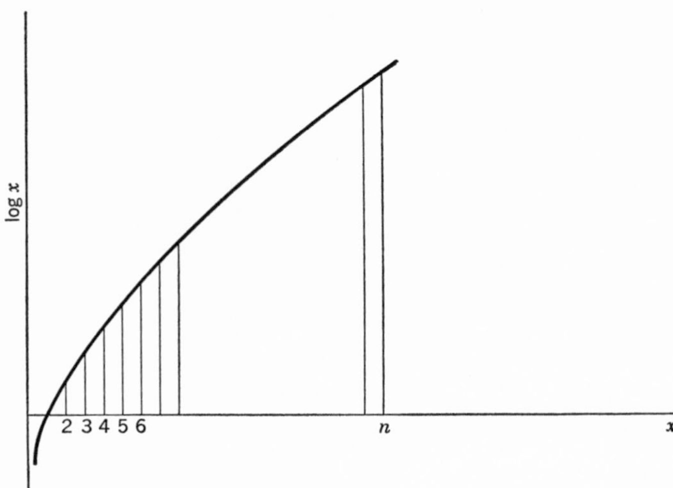

Fig. A1.1 The evaluation of log n!.

(ii) The second auxiliary result:

$$\lim_{n\to\infty}\left(1+\frac{1}{n}\right)^n = e$$

The expansion for $[1 + (1/n)]^n$ is

$$1 + \frac{n}{1!}\frac{1}{n} + \frac{n(n-1)}{2!}\left(\frac{1}{n}\right)^2 + \frac{n(n-1)(n-2)}{3!}\left(\frac{1}{n}\right)^3 + \cdots$$

As n becomes larger all the terms in n clearly tend to unity so that the series tends to

$$1 + \frac{1}{1!} + \frac{1}{2!} + \frac{1}{3!} + \cdots = e$$

as required.

We are now in a position to evaluate the expression (a1). Apply Stirling's theorem to the terms $(2n)!$, $(n + r)!$ and $(n - r)!$

$$(2n)! = (2n)^{2n}e^{-2n}\sqrt{2\pi \cdot 2n} = 2^{2n}n^{2n+(1/2)}e^{-2n}\sqrt{4\pi}$$

$$(n + r)! = (n + r)^{n+r}e^{-(n+r)}\sqrt{2\pi(n + r)}$$

$$= n^{n+r+(1/2)}\left(1 + \frac{r}{n}\right)^{n+r+(1/2)} e^{-n-r}\sqrt{2\pi}$$

$$(n - r)! = n^{n-r+(1/2)}\left(1 - \frac{r}{n}\right)^{n-r+(1/2)} e^{-n+r}\sqrt{2\pi}$$

Therefore,

$$(n + r)!(n - r)!$$

$$= n^{2n+1}\left(1 - \frac{r^2}{n^2}\right)^{n+(1/2)}\left(1 + \frac{r}{n}\right)^{r}\left(1 - \frac{r}{n}\right)^{-r} e^{-2n} \cdot 2\pi$$

The variable part of (a1) can now be written

$$\left(1 - \frac{r^2}{n^2}\right)^{-n-(1/2)}\left(1 - \frac{r}{n}\right)^{-r}\left(1 + \frac{r}{n}\right)^{r}$$

$$= \left(1 - \frac{r^2}{n^2}\right)^{-(n^2/r^2)(r^2/n)}\left(1 - \frac{r^2}{n^2}\right)^{-1/2}$$

$$\left(1 + \frac{r}{n}\right)^{n/r(-r^2/n)}\left(1 - \frac{r}{n}\right)^{-n/r(-r^2/n)}$$

Thus expression (a1) can now be written

$$\frac{1}{\sqrt{n\pi}}\left(1 - \frac{r^2}{n^2}\right)^{-(n^2/r^2)(r^2/n)}\left(1 - \frac{r^2}{n^2}\right)^{-1/2}$$

$$\left(1 + \frac{r}{n}\right)^{n/r(-r^2/n)}\left(1 - \frac{r}{n}\right)^{-n/r(-r^2/n)}$$

Now
$$\left(1 - \frac{r^2}{n^2}\right)^{-n^2/r^2} \to e \quad \text{as} \quad \frac{n}{r} \to \infty$$

$$\left(1 - \frac{r^2}{n^2}\right)^{-1/2} \to 1$$

$$\left(1 + \frac{r}{n}\right)^{n/r} \to e$$

$$\left(1 - \frac{r}{n}\right)^{-n/r} \to e$$

Thus finally, the probability of error R is

$$\frac{1}{\sqrt{\pi n}} e^{-r^2/n}$$

The significant feature of this result is the form e^{-r^2}. It specifies the probability of an error R and is thus equivalent to Equation (2.4) in which the error is the difference between the true value X and the measured value x. The only problem that remains in putting the equation into standard form is to redefine the constants. Put

$$hx = \frac{r}{\sqrt{n}}$$

for the exponent and in the constant replace $1/\sqrt{n}$ by $h\,dx$. The equation then reads

$$P(x)\,dx = \frac{h}{\sqrt{\pi}} e^{-h^2x^2}\,dx$$

where $P(x)\,dx$ is the probability of finding an error between x and $x + dx$.

A1.2 Standard Deviation of the Normal Distribution

We must calculate the sum of the squares of the errors divided by the total number of observations. Let there be N observations where N can be assumed to be a very large number.

The number of errors of magnitude between x and $x + dx$ equals $\frac{Nh}{\sqrt{\pi}} e^{-h^2x^2}\,dx$.

Therefore,
$$\sigma^2 = \frac{1}{N}\int_{-\infty}^{\infty} N\frac{h}{\sqrt{\pi}} e^{-h^2x^2} \cdot x^2\,dx$$
$$= \frac{h}{\sqrt{\pi}}\int_{-\infty}^{\infty} x^2 e^{-h^2x^2}\,dx$$

The integral is a standard one and has a value $\sqrt{\pi}/2h^3$

Therefore,
$$\sigma^2 = \frac{h}{\sqrt{\pi}} \cdot \frac{\sqrt{\pi}}{2h^3} = \frac{1}{2h^2}$$

This provides the justification for Equation (2.5) and enables us to rewrite the probability function

$$P(x)\,dx = \frac{1}{\sqrt{2\pi}\,\sigma} e^{-x^2/2\sigma^2}\,dx$$

A1.3 Areas Under the Normal Distribution Curve

The probability that an error falls between x and $x + dx$ is

$$\frac{1}{\sqrt{2\pi}\,\sigma} e^{-x^2/2\sigma}\,dx$$

Therefore, the probability that an error lies between 0 and x is

$$\int_0^x \frac{1}{\sqrt{2\pi}\,\sigma} e^{-x^2/2\sigma} dx$$

Although this integral can be easily evaluated for infinite limits it is not so simple for fixed limits as we now require.

x	Probability that an error lies between 0 and x
0	0
0.1	0.04
0.2	0.08
0.3	0.12
0.4	0.16
0.5	0.19
0.6	0.23
0.7	0.26
0.8	0.29
0.9	0.32
1.0	0.34
1.1	0.36
1.2	0.38
1.3	0.40
1.4	0.42
1.5	0.43
1.6	0.45
1.7	0.46
1.8	0.46
1.9	0.47
2.0	0.48
2.5	0.49
3.0	0.499

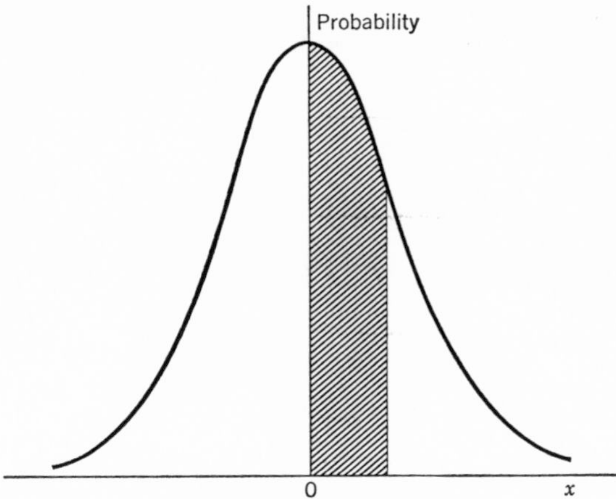

Fig. A1.2 The area evaluated in calculating the probability of occurrence of an error up to x.

Numerical methods of integration are used with results given in the table on page 184 (see Fig. A1.2).

If we require the probability that an error lies between $\pm x/\sigma$ the value is, of course, doubled. For example, the entry at $x/\sigma = 1$ is 0.34 giving the 68 per cent figure that we have been using for $\pm 1\sigma$ limits. The table is intended to give only an indication of the way the probabilities run and for statistical work reference should be made to one of the many statistical tables available (e.g., Reference 7 in the Bibliography).

2 Least Squares Fit to Straight Lines

Consider a set of observations (x_i, y_i) to which it is desired to fit a linear relation

$$y = mx + b$$

We assume that the x values are precise, that all the uncertainty is contained in the y values and that the weights of the y values are equal. The differences whose sum of squares it is desired to minimize are of the form

$$\delta y_i = y_i - (mx_i + b)$$

Therefore,

$$(\delta y_i)^2 = [y_i - (mx_i + b)]^2$$
$$= y_i^2 + m^2x_i^2 + b^2 + 2mx_ib - 2mx_iy_i - 2y_ib$$

If there are n pairs of observations the sum is

$$M = \Sigma\,(\delta y_i)^2 = \Sigma\, y_i^2 + m^2\,\Sigma\, x_i^2 + nb^2 + 2mb\,\Sigma\, x_i - 2m\,\Sigma\, x_iy_i - 2b\,\Sigma\, y_i$$

The condition for the best choice of m and b is that $\Sigma\,(\delta y_i)^2$ should be a minimum

$$\frac{\partial M}{\partial m} = 0 \quad \text{and} \quad \frac{\partial M}{\partial b} = 0$$

The first condition gives

$$2m\,\Sigma\, x_i^2 + 2b\,\Sigma\, x_i - 2\,\Sigma\,(x_iy_i) = 0$$

The second $\quad 2nb + 2m\,\Sigma\, x_i - 2\,\Sigma\, y_i = 0$

Solution of the simultaneous equations for m and b gives

$$m = \frac{n \Sigma (x_i y_i) - \Sigma x_i \Sigma y_i}{n \Sigma xt - (\Sigma x_i)^2}$$

$$b = \Sigma \frac{x_i^2 \Sigma y_i - \Sigma x_i \Sigma (x_i y_i)}{n \Sigma x_i^2 - (\Sigma x_i)^2}$$

Standard deviations for the m and b can be calculated as follows: The calculated values of m and b are functions of the quantities y_i. The standard deviations for m and b are, therefore, to be calculated using Equation (3.3) for the standard deviation of computed functions. They will be calculated in terms of the standard deviation of the y values. This was written as Equation (6.3) using the quantities δy_i

$$s_y = \sqrt{\frac{\Sigma (\delta y_i)^2}{n - 2}}$$

A justification of the value $n - 2$ will not be attempted but it is associated with the fact that the δy_i are not independent but are connected by the existence of the best line given by m and b.

Equation (3.3) for the standard deviation of a computed value says

$$s^2 = \left(\frac{\partial f}{\partial x}\right)^2 s_x{}^2 + \left(\frac{\partial f}{\partial y}\right)^2 s_y{}^2 + \cdots$$

This formula is applied to our case by noting that the x and y of the formula are the y_1, y_2, etc., which form part of our set of observations. Thus the function for m is

$$m = \frac{1}{n \Sigma x_i{}^2 - (\Sigma x_i)^2} [n x_1 y_1 - y_1 \Sigma x_i + n x_2 y_2 - y_2 \Sigma x_i + \cdots]$$

Therefore, $$\frac{\partial m}{\partial y_k} = \frac{1}{n \Sigma x_i^2 - (\Sigma x_i)^2} [nx_k - \Sigma x_i]$$

and

$$\left(\frac{\partial m}{\partial y_k}\right)^2 = \frac{1}{(n \Sigma x_i^2 - (\Sigma x_i)^2)^2} [n^2x_k^2 + (\Sigma x_i)^2 - 2nx_k \Sigma x_i]$$

Since s_y is common to all the contributions we can sum the $(\partial m/\partial k)^2$ directly to obtain

$$\sum_k \left(\frac{\partial m}{\partial y_k}\right)^2 = \frac{1}{(n \Sigma x_i^2 - (\Sigma x_i)^2)^2} [n^2 \Sigma x_i^2 + n(\Sigma x_i)^2 - 2n(\Sigma x_i)^2]$$

since $\Sigma x_k = \Sigma x_i$, etc.
Therefore,

$$\sum_k \left(\frac{\partial m}{\partial y_k}\right)^2 = \frac{1}{(n \Sigma x_i^2 - (\Sigma x_i)^2)^2} [n^2 \Sigma x_i^2 - n(\Sigma x_i)^2]$$

$$= \frac{n}{n \Sigma x_i^2 - (\Sigma x_i)^2}$$

or $$s_m = s_y \sqrt{\frac{n}{n \Sigma x_i^2 - (\Sigma x_i)^2}}$$

The value for s_b can be found by the same procedure.

BIBLIOGRAPHY

The following selection of books may prove useful in the further study of the topics mentioned in the text. It is not intended to be exhaustive.

1 Beers, Y., *Introduction to the Theory of Error*, Addison-Wesley, 1957. A short text dealing with the applications of error theory to experimenting.

2 Braddick, H. J. J., *The Physics of the Experimental Method*, Chapman and Hall, London, 1956. A study of practical experimenting techniques including observation treatment.

3 Cox, D. R., *Planning of Experiments*, Wiley, 1958. An account of the statistical methods for choosing an experiment program for maximum information (see page 94).

4 Fretter, W. B., *Introduction to Experimental Physics*, Prentice-Hall, 1954. An account of the experimental methods used in physics research.

5 Freund, J. E., *Modern Elementary Statistics*, Prentice-Hall, 1961. An introduction to statistics.

6 Jeffreys, H., *Scientific Inference*, Cambridge University Press, 1957. A study of the applications of probability theory to experimental information.

7 Lindley, D. V., and J. C. P. Miller, *Cambridge Elementary Statistical Tables*, Cambridge University Press, 1958.

8 Margenau, H., and G. M. Murphy, *The Mathematics of Physics and Chemistry*, Van Nostrand, 1947.

9 Menzel, D. H., H. M. Jones, and L. G. Boyd, *Writing a Technical Paper*, McGraw-Hill, 1961. A survey of the principles and practice of composition in scientific writing.

10 Parratt, L. G., *Probability and Experimental Errors in Science*, Wiley, 1961. A study of the applications of probability and statistics in measurement.

11 Schenck, H., *Theories of Engineering Experimentation*, McGraw-Hill, 1961. An account of those experiment design techniques which refer more specifically to engineering systems.

12 Shamos, M. H. (editor), *Great Experiments in Physics*, Holt-Dryden, 1960. A compilation of the writings of great physicists describing their own experiments.

13 Stanton, R. G., *Numerical Methods for Science and Engineering*, Prentice-Hall, 1961. This text is devoted mostly to numerical methods in mathematics but many of the techniques have application to curve fitting.

14 Strong, J., *Procedures in Experimental Physics*, Prentice-Hall, 1938. An account of the experimental methods used in physics research.

15 Topping, J., *Errors of Observation and their Treatment*, The Institute of Physics, London, 1955. A short text dealing with the applications of error theory to experimenting.

16 Tuttle, L., and J. Satterley, *The Theory of Measurements*, Longmans, Green, 1925. An account of the application of statistics to measurement and of numerical methods in the treatment of observations.

17 Whittaker, E. T., and G. Robinson, *The Calculus of Observations*, Blackie, Glasgow, 1944. An extensive treatment of the mathematical aspects of observation treatment.

18 Wilson, E. B., *An Introduction to Scientific Research*, McGraw-Hill, 1952. An account of the nature of scientific experimenting and the methods used.

19 Wortham, A. W., and T. E. Smith, *Practical Statistics in Experimental Design*, Charles E. Merrill, 1959. A short treatment of the experiment design methods used in engineering.

20 Worthing, A. G., and J. Geffner, *Treatment of Experimental Data*, Wiley, 1946. An account of the application of statistics to measurement and of numerical methods in the treatment of observations.

PROBLEM SOLUTIONS

Chapter 2

1.

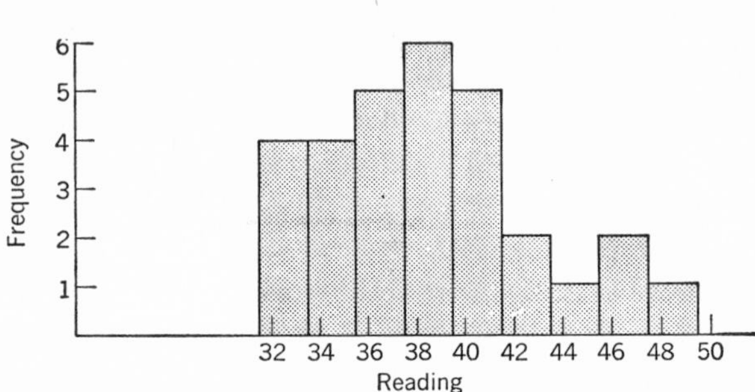

2. Between 38 and 39; 38
3. 38.30
4. 4.388
5. 0.801
6. 0.576
7. a) ± 4.388 about the universe mean
   b) ± 8.776 about the universe mean
8. a) ± 0.801 about the universe mean
   b) ± 1.602 about the universe mean
9. a) ± 0.576 about the universe standard deviation
   b) ± 1.152 about the universe standard deviation
10. $h = 0.161$, $p = 2.940$
11. rejection
13. about 130
14. about 207

Chapter 3

1. ± 0.05 cm; 0.4%
2. 0.5%
3. 5 cm; 1 cm
4. a) No; b) Yes
5. 0.007%
6. a) 1%; b) 5%
7. a) 2 sec; b) 100 sec
8. 0.07% assuming the watch gains uniformly
9. 3.7%

Chapter 3 (continued)

10. 0.2 ohms
11. 0.012 g cc^{-1}
12. 5.5%
13. 1.2×10^{-11} dynes cm^{-2}
14. 0.25%
15. 0.05%
16. 77 ohms
17. 1.9 cals g^{-1}
18. approximately 50°
19. 0.2 ohms
20. 3.6°
21. 0.005
22. 7.6% the wrong way; 3.2% the right way
23. 0.00015
24. 0.045
25. 2.1
26. 0.032
27. 14.9 cm sec^{-2}
28. 0.0027 cgs units

Chapter 5

1. No
2. range $\propto \left(\dfrac{\text{velocity}}{g}\right)^2$
3. pressure $\propto \dfrac{\text{surface tension}}{\text{radius}}$
4. period $\propto \sqrt{\dfrac{\text{moment of inertia}}{\text{rigidity constant}}}$
5. deflection $\propto \left(\dfrac{\text{load force}}{y \times \text{radius}^2}\right)^a \times \left(\dfrac{\text{length}}{\text{radius}}\right)^b$ radius where a and b are arbitrary constants
6. s vertically, t^2 horizontally, slope is $\frac{1}{2}a$
7. T vertically, n^2l^2 horizontally, slope is 4 m
8. P vertically, v^2 horizontally, slope is $\rho/2$
9. T^2 vertically, $\cos\alpha$ horizontally, slope is $4\pi^2 l/g$
10. d vertically, Wl^3 horizontally, slope is $4/Yab^3$
11. h vertically, $1/R$ horizontally, slope is $2\sigma/\rho g$
12. p vertically, T horizontally, slope is R/v
13. fv_0 vertically, $f - f_0$ horizontally, slope is v
14. l vertically, Δt horizontally, slope is $l_0\alpha$ and intercept is l_0, whence α
15. $\sin\theta_1$ vertically, $\sin\theta_2$ horizontally, slope is μ_2/μ_1
16. $1/s$ vertically, $1/s'$ horizontally, each intercept is $1/f$; or ss' vertically, $s + s'$ horizontally, slope is f
17. $1/c$ vertically, ω^2 horizontally, slope is L.

Chapter 5 (continued)

18. F vertically, $1/r^2$ horizontally, check for linearity

19. F vertically, $i_1 i_2/r^2$ horizontally, check for linearity and check F vs. i_1, F vs. i_2 and F vs. $1/r$ separately, holding other variables constant.

20. $\log Q$ vertically, t horizontally, slope is $-1/RC$

21. Z^2 vertically, $1/\omega^2$ horizontally, slope is $1/c^2$ and intercept is R^2

22. m^2 vertically, m^2v^2 horizontally, slope is $1/c^2$ and intercept is m_0^2

23. $1/\lambda$ vertically, $1/n^2$ horizontally, slope is $-R$, intercept is $R/4$.

24. $\log J/T^2$ vertically, $1/T$ horizontally, slope is $-\varphi/k$, intercept is $\log A$

25. $m_w(T_2 - T_3)$ vertically, $m_s(T_1 - T_2)$ horizontally, slope is S.

26. $m(T_4 - T_3)$ vertically, $T_2 - T_1$ horizontally, slope is KA/d

27. $VI/(\frac{1}{2}(T_1 + T_2) - T_0)$ vertically, $m(T_2 - T_1)/(\frac{1}{2}(T_1 + T_2) - T_0)$ horizontally slope is JS.

Chapter 6

1. 14.2 ± 0.1

2. 14.25 ± 0.15

3. 6.75 ± 0.03

4. 38.30 ± 0.80

5. c) 12.5×10^{-4} ohm² sec²
   d) 5.63×10^{-3} henry
   e) 11.2×10^{-4} and 14.2×10^{-4} ohm² sec²; $+1.7 \times 10^{-4}$, -1.3×10^{-4} ohm² sec²
   f) 5.33×10^{-3} henry to 6.00×10^{-3} henry gives a maximum range of uncertainty of $\pm 0.37 \times 10^{-3}$ henry
   g) 6.13
   h) 5.4 and 6.7 are the limits giving a maximum range of uncertainty of ± 0.7
   i) $L = (5.6 \pm .4) \times 10^{-3}$ henry
   $R = 6.1 \pm 0.7$ ohm

6. Mean is 17.44 and standard deviation of mean is 0.25

7. a) Slope $= 4.988 \times 10^{-2}$ ohms deg^{-1}; intercept $= 11.916$ ohms
   b) 4.186×10^{-3} deg^{-1}
   c) $s_{\text{slope}} = 2.430 \times 10^{-3}$ ohms deg^{-1}
   $s_{\text{intercept}} = 0.1227$ ohms

Chapter 6 (continued)

d) $s_\alpha = 2.084 \times 10^{-4}$ deg^{-1}

e) $\alpha = 4.19 \times 10^{-3} \pm 0.21 \times 10^{-3}$ deg^{-1}

$Ro = 11.92 \pm 0.12$ ohms

8. a) $i = 0.5\, e^{2v}$

b) $y = 0.6\, x^{2.4}$

c) $f = 6.2\, e^{-365/T}$

Index